THE
PLANETS

ANDREW COHEN

WITH

PROFESSOR BRIAN COX

THE

PLANETS

WILLIAM
COLLINS

TO ANNA – AMONGST THE VASTNESS OF THIS STORY
HOW LUCKY AM I TO HAVE FOUND YOU.
ANDREW COHEN

William Collins
An imprint of HarperCollins*Publishers*
1 London Bridge Street
London SE1 9GF

WilliamCollinsBooks.com

First published in Great Britain by William Collins in 2019

2022 2021 2020 2019
10 9 8 7 6 5 4 3 2 1

Text © Brian Cox & Andrew Cohen 2019
Images © Individual copyright holders
Diagrams © HarperCollins*Publishers* 2019

By arrangement with the BBC

The BBC logo is a trademark of the British Broadcasting
Corporation and is used under licence

BBC logo © BBC 2014

Brian Cox and Andrew Cohen assert the moral right to
be identified as the authors of this work in accordance
with the Copyright, Designs and Patents Act 1988

A catalogue record for this book is available from the
British Library

ISBN 978-0-00-748884-1

Design: Zoë Bather
Editor: Helena Caldon
Managing Editor: Hazel Eriksson
Picture Research: Madeleine Penny
Scientific Consultant for William Collins: Ian Ridpath
Colour Processing/Production: Tom Cabot/ketchup

Printed in Spain by Graficas Estella

This book is produced from independently certified
FSC™ paper to ensure responsible forest management.
For more information visit:
www.harpercollins.co.uk/green

ACKNOWLEDGEMENTS

It is a true privilege to work with world-class
people on a project like *The Planets* and
from start to finish the television series that
accompanies this book has been driven by the
most talented of teams – this is a small chance
to say a big thank you to you all.

Gideon Bradshaw who we have had the
pleasure of working with for so many years
has led the team with his normal combination
of immense creativity and calm leadership,
finding solutions when it often looked like we'd
run out of ideas.

Working with him has been a world-class
team of film-makers with Stephen Cooter
leading the way and setting out a vision for the
series not just on paper but on location and in
the edit as well. We were very lucky to have
Martin Johnson and Nic Stacey to complete
the directing team on *The Planets*, producing
a beautiful set of films that both stand alone as
exceptional examples of science film-making
and together as a cohesive and interconnected
series. Not an easy task.

They were supported by a hugely talented
team who have grappled with an endless array
of challenges that have been overcome in the
most creative of ways.

So a very big thank you to Zoe Heron, Katy
Savage, Kimberley Bartholomew, Victoria
Weaver, Poppy Pinnock, Graeme Dawson,
Louise Salkow, Ged Murphy, Paul Crosby,
Tom Hayward, Julius Brighton, Andy Paddon,
Chris Youle Grayling, Olly Meacock, Sophie
Chapman, Emma Chapman, Toby Keeley,
Kate Moore, John Gillespie, Maggie Oakley,
Krissi Loppas, Miguel Arnott Muzaly, Sarah
Houlton, Vicky Edgar and Marie O'Donnell.

We'd also like to thank Nicola Cook, Nik
Sopwith and Josh Green for the brilliant
initial development work that allowed us to
make *The Planets*.

We'd also like to particularly thank Laura
Davey who has as always been utterly crucial
to the success of the whole project and Evih
Efue who has skilfully managed to keep us
moving forward when everything had been
threatening go off track.

A very special thank you to the team at Lola
Post Production – Rob Harvey and his team
have brought an extraordinary visual scale to
the series, partnering us every step of the way.

We'd also like to thank our consultants
from The Open University Professor Stephen
Lewis and Professor David Rothery, their
support and advice has been invaluable on this
most complex of projects.

Finally we'd like to say thank you to the
truly brilliant team at William Collins. You
have produced the most beautiful of books
and we are lucky to work with such talent.
So thank you to Zoë, Helena and Hazel.
And of course a very big thank you to Myles
Archibald – his appetite for root canal
treatment shows no sign of waning and yet
through all of the pain he never ceases to at
least attempt a smile.

006

SOLAR SYSTEM

AN INTRODUCTION

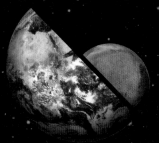

064

EARTH
+
MARS

THE TWO SISTERS

018

MERCURY
+
VENUS

A MOMENT IN THE SUN

118

JUPITER

THE GODFATHER

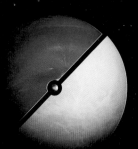

168

SATURN

THE CELESTIAL JEWEL

230

URANUS
NEPTUNE
PLUTO

INTO THE DARKNESS

282

INDEX

AN INTRODUCTION

SOLAR SYSTEM

PROFESSOR BRIAN COX

'Our Solar System is a wild frontier.'
Carrie Nugent

'Imagination will often carry us to worlds that
never were. But without it we go nowhere.'
Carl Sagan

WANDERING LIGHTS

In the daytime, our universe stretches only as far as the horizon. The Sun hides in plain sight because it is too bright for us to see it directly. Only rarely do we glimpse a watercolour moon. Unless we think hard, our intellects are confined to the surface of the Earth. After sunset, beyond cities, the Universe appears; a destination for the imagination, albeit separated by a seemingly unbridgeable gulf. This may be true for the stars, but it is not so for the planets. There are times when Mars, Venus, Jupiter and Saturn dominate the sky; bright lights that shift position nightly against the fixed stars, commanding our attention even if we aren't certain what we're looking at. The distances are still vast by terrestrial standards, but despite appearances the gulf is certainly not unbridgeable, because we have visited all of these planets and taken our first steps into the outer reaches of the Solar System beyond. And yet the wandering lights in the dark still feel detached from human affairs, and the time and effort we've spent in visiting them might seem to be an indulgence. This assumption, however, is profoundly wrong.

Below: The Milky Way in the night sky over Sliding Spring Observatory, New South Wales, Australia.

The exploration of the planets is not an indulgence. If we want to know how we came to be here we need to understand the histories of the planet that gave birth to us and the system that gave birth to it. We are children of Earth and also children of the Solar System.

The Solar System is a system. The Sun and the eight major planets and countless billions of minor planets, moons, asteroids, comets and unclassified lumps of ice and rock were formed back in the mists of time and they continue to evolve as one. We rarely notice the dynamic, interconnected nature of our system, although asteroid strikes on our planet are not such a rare occurrence. The Chelyabinsk impact in February 2013 injured 1,500 people when a 12,000-tonne asteroid broke up as it entered the Earth's atmosphere at 60 times the speed of sound, and the Tunguska airburst in Siberia in 1908 flattened 800 square miles of forest in an explosion comparable to that of the most powerful hydrogen bomb ever tested. The surface of the Moon bears testament to a record of violence and destruction from the skies that the Earth has also endured, but the relentless erasure of craters by weathering and our good fortune that no major impacts have occurred in recorded human history are the reason for our misplaced sense of isolation from the heavens.

Below: Meteor showers are not uncommon, but as most meteors are smaller than sand grains, they disintegrate before hitting the Earth's surface.

Right: The Tunguska airburst of 1908 is the largest impact event on Earth in recorded history. It flattened 800 square miles of forest.

The interdependent nature of the Solar System has become more evident as we have begun to understand its history. It is tempting to imagine that the physical layout of the planets is a fossilised remnant of primordial patterns in the collapsing dust cloud around the newly ignited Sun 4.6 billion years ago, but our exploration of the planets, coupled with increasingly powerful computer simulations of the evolution of the Solar System, has revealed that this is not the case. Planetary orbits are prone to instability; particularly so in the early, more chaotic years when our Solar System was young. The details of precisely how the planetary orbits have shifted are still uncertain, but we now suspect that Mercury, the innermost planet, began life much further out and was deflected inwards to its present-day seared orbit. Jupiter and Saturn may have drifted inwards shortly after their formation, before reversing their course and retreating, but not before affecting the distribution of material out of which Mars and Earth would later form. Around the time that life began on Earth, Neptune and Uranus may have been flung outwards, disrupting the orbits of billions of smaller objects far from the Sun. The record of this time of unprecedented violence, known as the Late Heavy Bombardment, is written across the scarred surface of the Moon, itself most likely formed in a glancing interplanetary collision between Earth and a Mars-sized planet 4.5 billion years ago. The planets are like snowflakes; the detail of their structure – their composition, size, spin and climate – are a frozen record of their past.

An understanding of the planets beyond Earth is therefore a prerequisite for understanding our home world, and that in turn is a prerequisite for understanding ourselves. Earth is unique in the Solar System because it is a planet with a complex ecosystem. The genesis and subsequent 4-billion-year evolution of life on Earth required planetary characteristics which are necessarily linked to the evolution of the system as a whole. There had to be liquid water on the surface, and much of this water was delivered after the Earth's formation by icy, water-rich asteroids and comets, possibly deflected inwards from the outer Solar System by Jupiter. These rivers, seas and lakes of extra-terrestrial origin had to persist for the best part of 4 billion years, which required a stable atmosphere to maintain surface temperatures and pressures within a limited range. Four billion years is a long time – around a third of the age of the Universe. The Sun has brightened by 25 per cent since the Earth formed, which makes the stability of our environment all the more difficult to understand. In a chaotic system of planets around an evolving star, a planet with life-supporting properties and the remarkable stability enjoyed by Earth over billions of years may be extremely unusual. The study of our sister worlds, Mars and Venus, has proved instructive in understanding just how fortunate we may have been and how delicate our position today might be.

Above: Two views of the Chelyabinsk meteor fireball caught on camera in 2013.

Above: Satellite image of the Earth, centred on the Pacific Ocean. Water dominates this hemisphere of the Blue Planet.

Opposite: The far side of the Moon bears the scars of the Late Heavy Bombardment. Photographed from Apollo 16 in 1972.

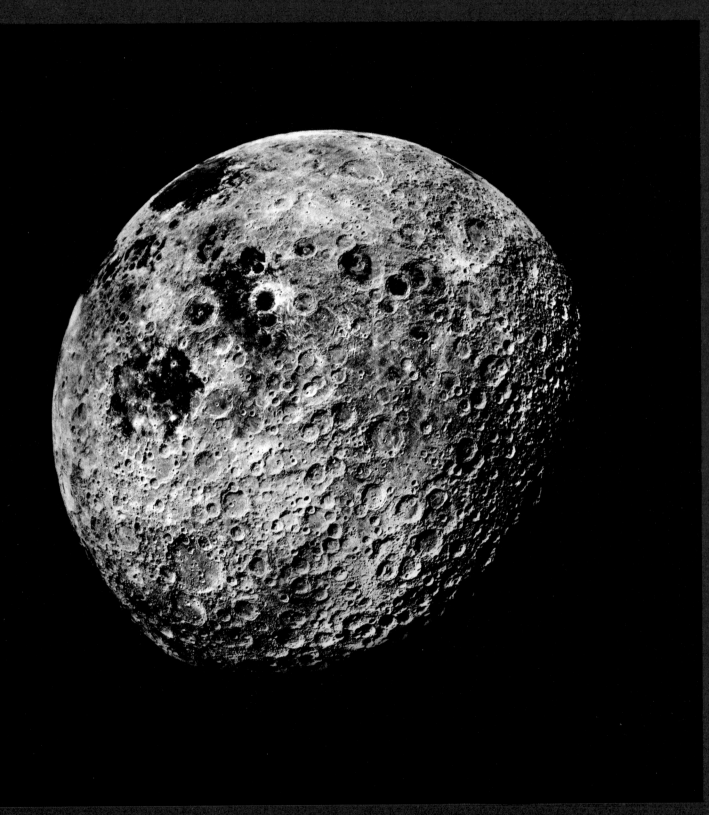

Four billion years ago, as life began on Earth, Mars was also Earth-like.

Four billion years ago, as life began on Earth, Mars was also Earth-like. It had oceans and rivers and active geology and complex surface chemistry; the ingredients of life. One of the primary goals of the fleet of spacecraft currently in orbit around and exploring the surface of Mars is to search for evidence of past or even present life, and to understand why the red planet was transformed from a potential Eden at the dawn of the Solar System to the frigid desert world we observe today. The story is complex, but one of the most important differences between the two worlds is size. Mars is just one-tenth the mass of Earth; too small to hang on to its internal heat, its protective magnetic field and its atmosphere for much more than a billion years after its formation. Yet Mars formed in a similar region of the Solar System to Earth and Venus, so why is it so small? The answer may lie in the fast-changing orbits of Jupiter and Saturn early in the Solar System's history. These surprising findings will be explored later in this book.

Opposite: This colour-enhanced satellite photo of the Mississippi River Delta shows a lush, watery landscape.

Above: NASA's Mars Reconnaissance Orbiter captured this photograph of Aram Chaos, an ancient impact crater that once held a lake.

'Life, forever dying to be born
afresh, forever young and eager, will
presently stand upon this Earth as
upon a footstool, and stretch out its
realm amidst the stars.'
H.G. Wells

The history of Venus is perhaps even more puzzling, in part because of the immense difficulty of exploring the planet. Venus is often described as a vision of hell; surface temperatures are high enough to melt lead, and the atmospheric pressure is 90 times that on Earth. Sulphuric acid raindrops fall from its clouds. And yet, long ago, Venus too may have been Earth-like. Perhaps there were once Venusians, before a runaway greenhouse effect took hold and began destroying Venus's temperate climate around 2.5 billion years ago – although this date is highly uncertain.

Taken together, the stories of the three large terrestrial planets are salutary. If an alien astronomer observed our Solar System from afar, they would classify Mars, Earth and Venus as potentially living worlds, orbiting as they do inside the so-called habitable zone around the Sun – the region within which, if atmospheric conditions are right, liquid water can exist on the surface of the planets. All three worlds may have once been habitable, and all three worlds may have once harboured life, but now only Earth supports a complex ecosystem, let alone a civilisation.

Understanding why Mars and Venus diverged so significantly from Earth over the last 4 billion years will provide great insight into the fragility of worlds and perhaps suggest whether our own good fortune is near-impossible to comprehend or merely outrageous. Planets change. Ours could change at any moment. A stray comet from the frozen Kuiper Belt beyond the orbit of Neptune could put an end to our story. We could also put an end to ourselves. The study of Venus might help us avoid one of the ways by which we could destroy our civilisation, because it shows us what greenhouse gases can do to a world. I think one of the reasons why anthropogenic climate change is so difficult for a certain type of person to accept is that atmospheres seem ethereal and tenuous and incapable of trapping enough heat to modify the temperatures on a planet significantly. For such people I suggest a trip to Venus, where they will be squashed and boiled and dissolved on the surface of Earth's twin.

The exploration of the planets, then, is not an indulgence. If we want to know how we came to be here we need to understand the histories of the planet that gave birth to us and the system that gave birth to it. We are children of Earth and also children of the Solar System. Understanding our history is important because it places our existence in context. The more we learn about the events that led to the emergence of humans on this planet only a few hundred thousand years ago, the more we are forced to marvel at the sheer unlikeliness of it all. We needed Jupiter and the comets and asteroids and countless collisions and mergers and near-catastrophes stretching back 4.6 billion years. There are valid objections to this way of thinking; it is an objective fact that we are here, and our future should be our primary concern. That may be so, but I argue that a deeper understanding of the evolution of the planets is essential for our continued prosperity and existence on this one. The threat of catastrophic climate change is an obvious example, but there are many other reasons why knowledge is important. There is feedback in human affairs; our collective state of mind affects the decisions we make. To confine our imaginations to the surface of the Earth is to ignore both our immense good fortune and the fragility of our position. A wider knowledge of both will, I believe, help secure a safer and more prosperous future.

Opposite: The Apollo 11 mission
in July 1969 changed human
history, landing the first people
on the surface of the Moon.

MERCURY

+

VENUS

A MOMENT IN THE SUN

ANDREW COHEN

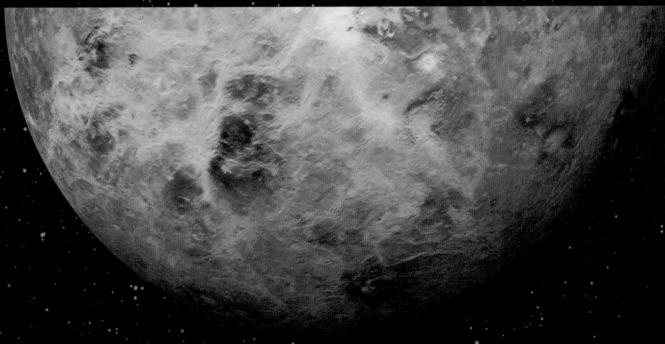

'Life's but a walking shadow.'
Macbeth, William Shakespeare

'A runaway greenhouse effect, which is what
happened to Venus, is a perfect example of
what could happen to Earth in the future if
we don't do anything about it.'
Lynn Rothschild

DAYS YET TO COME

'What has been is what will be, and what has been done is what will be done, and there is nothing new under the sun.'
Ecclesiastes 1:9

Earth sits a mere 150 million kilometres from the Sun – not too hot, not too cold, with surface temperatures ranging from minus 88 to plus 58 degrees Celsius. This 'Goldilocks' location has created a stability of climate that, despite the best efforts of ice ages and impacts, has allowed life to maintain an unbroken chain for nearly 4 billion years, and yet we know for certain that it cannot last.

Our Sun, like every star in the universe, is far from static. Stars have life cycles of their own and, eventually, the hydrogen fuel that powers the nuclear reactions within a star will begin to run out and the star will enter the final phases of its lifetime. It will expand, cool and change colour to become a red giant. Small stars, like our sun, will undergo a relatively peaceful and beautiful death, which will see it pass through a planetary nebula phase to become a white dwarf, which will cool down over time to leave a brown dwarf. Life on Earth has prospered through our sun's middle years, but these optimum conditions are waning. At first the changes will be invisible, but a billion years from now they will be obvious to any life forms left on the planet – an immense sun, filling the sky, will warm and transform itself and the Earth that it shines upon. The Sun is both the giver and the taker of life on our planet.

It is one of the great paradoxes of the Universe that as the life of a star like ours begins to wane, its size and luminosity will increase. A rise in luminosity of just 10 per cent will see the average surface temperature on Earth rise to 47 degrees Celsius instead of the 15 degrees Celsius that it is today. The effect of this rise in temperature manifests in a lifting of vast amounts of water vapour from the oceans into the atmosphere, creating a greenhouse effect that could quickly and rapidly run out of control, evaporating the oceans and sending the surface temperature skyrocketing. Astrobiologist David Grinspoon explains,

> *The greenhouse effect is the name we give to the physical process by which planets heat up through the interaction of their atmospheres and solar radiation. Solar radiation comes in what we call the visible wave lengths, primarily wave lengths that we can see, and most atmospheric gases are very transparent to visible radiation. So light from the Sun comes through pretty much unimpeded by an atmosphere and reaches the surface of a planet. Then the surface of the planet reradiates that radiation in infrared, because planets are much cooler than the Sun. And that means they radiate at much longer wave lengths – what we call infrared. That infrared radiation doesn't make it through an atmosphere so easily. Some of the atmospheric gases, the ones we call greenhouse gases, block infrared radiation and so therefore the more of those greenhouse gases that are in a planet's atmosphere the harder it is for that surface radiation to make it back out into space and the more that planet will heat up.*

Estimates of the timescale that will see our oceans disappear vary massively, and are heavily influenced by a multitude of factors, but few are in doubt that by the time our planet reaches its 8-billionth birthday (in 3.5 billion years' time) the end will be in sight. With temperatures heading above a thousand degrees, life will have long disappeared from a surface that is beginning to melt under the burning Sun.

Above and opposite: These computer artworks show how the Earth might appear in 5–7 billion years' time. As the Sun swells and becomes a red giant, temperatures on our planet's surface will soar, making life untenable.

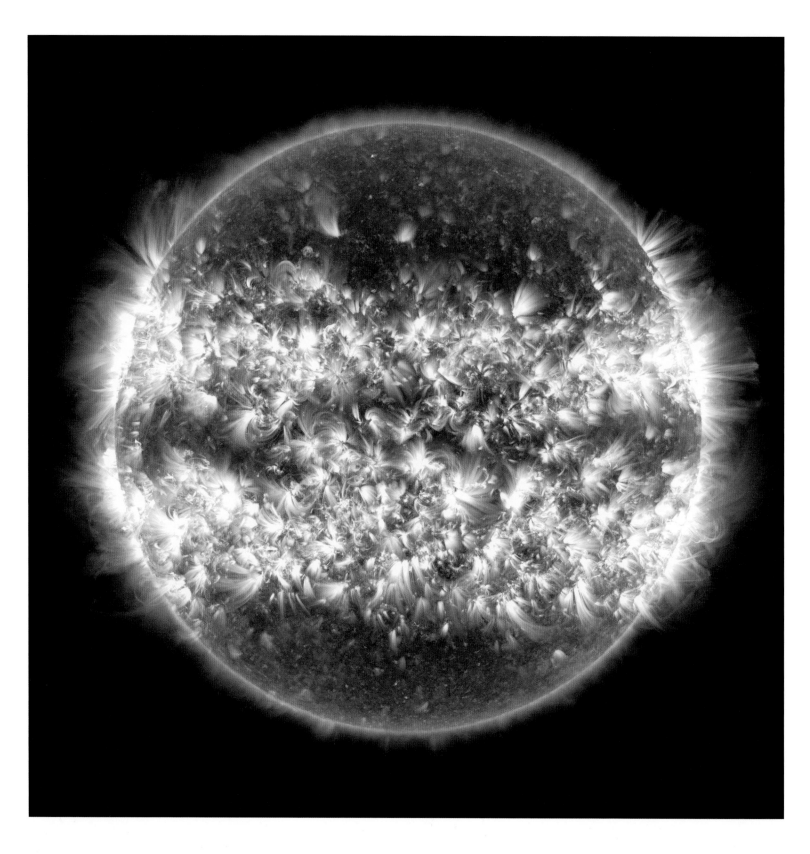

Above: Our Sun is far from static, and NASA's Solar Dynamics Observatory regularly and consistently tracks its rise to solar maximum. This composite image shows 25 shots taken between 16 April 2012 and 15 April 2013, which reveal an increase in solar activity.

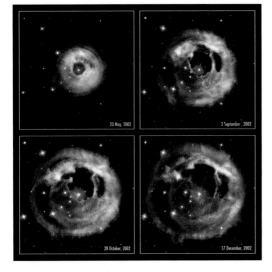

Astrobiologist David Grinspoon on the greenhouse effect

'The greenhouse effect gets a bad rap these days and that's understandable because we are tweaking it in ways we don't fully understand, in changing the climate balance that we depend upon on this planet.

'And yet it's important to understand that the greenhouse effect is an essential part of what makes the Earth a habitable planet. Without some measure of greenhouse effect, Earth would be completely frozen over and life would not be possible on this planet.

'Thirty degrees or so of a greenhouse effect on a planet like the Earth is absolutely wonderful and crucial and what keeps us alive, what makes Earth such a great place for life because it keeps us in that liquid water zone.

'But, of course you can have too much of a good thing, look at Venus for a possible image of Earth's future, if we're not careful.'

Moving even further into the future the outlook becomes bleaker. As the Sun enters old age it will grow into a red giant, engulfing the Earth within its expanding atmosphere. Moonless, lifeless and perhaps reduced to its inner core, our planet and the civilisations it once harboured will be nothing but a distant memory, etched in the atoms that made us all as they are dispersed amongst the cosmos.

For planet Earth, the clock is ticking and time is slowly running out, but ours is far from the only world to enjoy its moment in the sun. Across the history of our Solar System, stretching deep into its ancient past and reaching far into its future, we see stories of worlds in a constant battle with our ever-changing star. Close in, ancient worlds such as Mercury, which long ago lost their fight with the Sun, are taunted by views of Earth, and what might have been. Further out and even hotter, Venus circles, shrouded in a choking cloak of cloud, and even further beyond the Earth, Mars sits cold and barren. Beyond these planets, frozen worlds await, huddled in perpetual hibernation; anticipating the moment when the warmth of the Sun reaches out far enough, with sufficient heat, to trigger a first spring. On that day, mountains of ice will melt, rivers of water will flow, and where there was once only bleakness, in the distant future on planets once frozen and lifeless we might find a place that looks very much like home.

The story of our Solar System is not as we once thought it – eternal and unchanging. Instead it is a place of endless transformation. It is a narrative that repeats itself with a predictable rhythm, and as one world passes, another comes into the light. Only one planet has maintained stability for almost the entire life of the Solar System: the Earth has remained habitable for at least 4 billion years while change has played out all around it. What makes the Earth so lucky compared to all of its terrestrial siblings? To answer that question we need to look not just at our planet but at the whole of the Solar System, going right back to the very beginning.

Above: This series of photos, captured by the Hubble Space Telescope in 2002, demonstrates the reverberation of light through space. A burst of light from an unusual star in the constellation spreads through space and reflects off surrounding dust. During this activity, the red star at the centre brightens to more than 600,000 times the Sun's luminosity. It will continue to expand before eventually disappearing.

Right: The last colourful hurrah of a star. Ultraviolet light from the dying star causes a glow around the white dwarf at the centre, where the star has burned out. This planetary nebula tells the story of the demise of our own Sun.

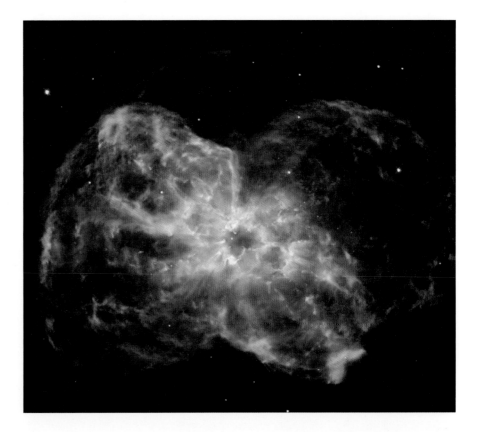

IN THE BEGINNING

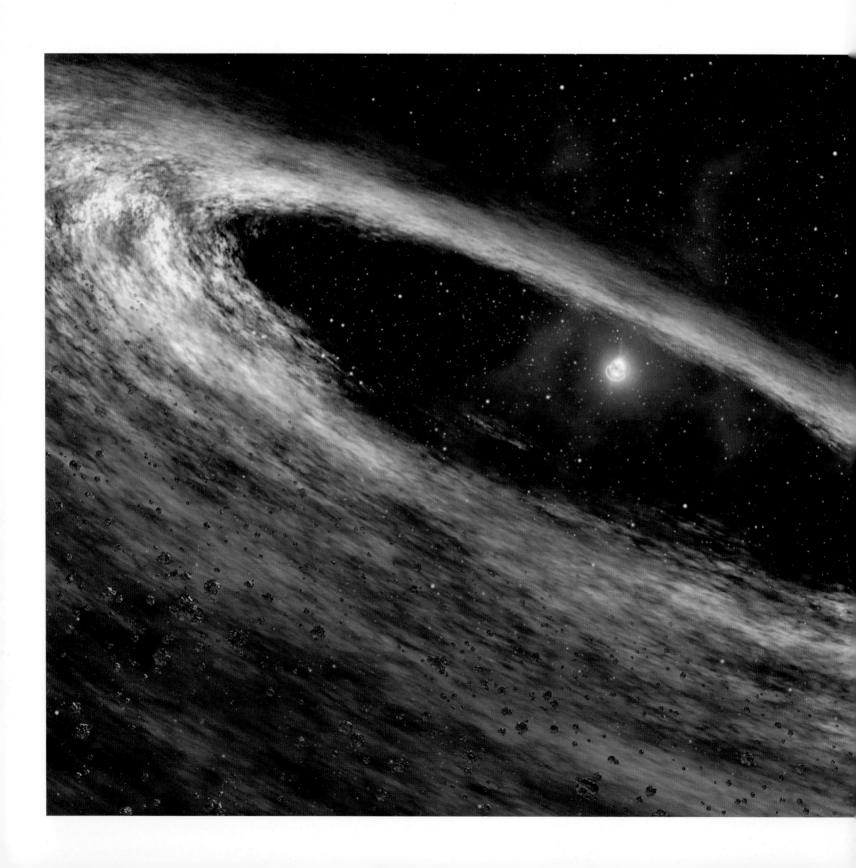

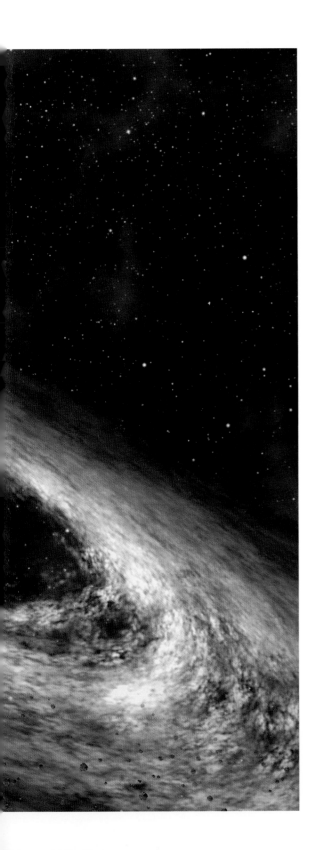

Below and diagram: This artist's image represents a dead star known as a pulsar. The disc of rubble that surrounds it resembles the protoplanetary discs of gas and dust that are found around young stars, which collide and coalesce under the gravitational force of attraction to create young planets.

For the first few million years after its birth, there were no terrestrial worlds to see the Sun rise, and there were no days, no nights, no circular tracks around it. Instead, surrounding our infant star was a vast cloud of dust and gas. A tiny fraction of the material left over from the Sun's formation, this swirling cloud would one day coalesce to form the various planets of the Solar System, and many other smaller bodies, but at this time, 4.7 billion years ago, there was nothing but tiny specks of dust reflecting back the light of our slowly growing star.

Only time – vast amounts of empty time – would allow enough of this gas and dust to catch and cluster, randomly forming the smallest of seeds. Most of these seeds would hardly get the chance to grow at all, smashed apart and returned to the immense swirl of dust from whence they arose. Just a few would grow big enough and survive long enough to capture and condense more of the cloud, slowly increasing their mass and density.

We still do not fully understand the process by which grains of dust no thicker than a human hair can amass to become rocky objects the size of a car, and as yet no model exists to explain this part of the evolution of a planet. But what we do know is that once that disc of gas and dust becomes populated with clumps of rock that make it past the 'metre-size barrier', a powerful force comes into play to propel the process forward. These newly formed planetesimals are big enough to allow the great sculpting force of gravity to draw the clumps together, growing to sizes of over a kilometre in length. Swirling around the Sun, thousands upon thousands of these objects live and die, colliding and coalescing under the increasing gravitational forces of attraction until eventually just a few emerge as planetary embryos, moon-sized bodies known as protoplanets. In the last violent steps of the process of planetary birth these protoplanets swirl around in crowded orbits, and many are destroyed, returned to the dust of their origins, but occasionally when a collision brings two or more of these giant objects together, the size of this mass of rock becomes big enough for gravity to pull it in from all sides, creating a sphere of newly formed rock, a new world. In that moment a planet is born.

Each of the terrestrial planets in our Solar System was born this way. They are the survivors of a process that destroys far more worlds than it ever creates, and which left just four rocky planets remaining – starting closest to the Sun with Mercury, then Venus, Earth and finally, farthest out, the cold and dead world of Mars. Today these four worlds all look vastly different, and yet all were created the same way, made up of the same ingredients and orbiting the same star. So why have they ended up so distinct from each other, and with such starkly different environments? And what makes this place, the Earth, so unique, the only one of the rocks that has blossomed with life? To understand, we have to look deep into the past of our Solar System, to explore the unique history of each of the planets by means of amazing feats of human engineering across billions of miles, and into environments of unknown and unimaginable extremes.

GROWING A ROCKY PLANET

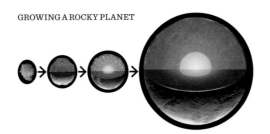

EXPLORING MERCURY

Getting to the smallest planet in the Solar System is anything but easy. Skirting past the Sun at a distance of just 46 million kilometres at the closest point in its orbit, Mercury is a planet that is not only held deep in the gravitational grip of our massive star but is also moving at an average orbital speed of 48 kilometres per second (km/s), by far the fastest-orbiting planet in the Solar System and far outpacing the Earth's more leisurely 30 km/s. It needs to whip around this quickly, otherwise it would have fallen into the Sun's embrace long ago, but the combination of its speed and position make it a planet that's immensely difficult to get close to, and even harder to get into orbit around. In order to do so, you have to travel fast enough to catch up with Mercury but not so fast that you cannot somehow slow down to prevent a headlong descent into the Sun, and that challenge has meant that until relatively recently it was the least explored of all the terrestrial planets.

For many decades our first and only close-up glimpse of the innermost rock orbiting the Sun came from the Mariner 10 spacecraft, when on three separate occasions in 1974 and 1975 it briefly flew past Mercury. This was the first spacecraft to use another planet to slingshot itself into a different flightpath, using a flyby of Venus to bend its trajectory to allow it to enter an orbit that would bring it near enough to Mercury to photograph it close up. Clad in protection to ensure it could survive the intense solar radiation and immense extremes of temperature, Mariner 10 was able to send back the first detailed images of Mercury as it flew past at just over 200 miles above its surface. It passed by the same sunlit side of Mercury each time, so it was only able to map 40 to 45 per cent of Mercury's surface.

The spacecraft took over 2,800 photos, which gave us never-before-seen views of the planet's cratered, Moon-like surface, a surface that we had never previously been able to fully resolve through Earth-based observation. Despite the beauty of the pictures taken, it wasn't the images from Mariner 10 that really surprised us, it was the data the probe collected relating to Mercury's geology, which pointed to a much more surprising history of the planet than had previously been imagined. Mercury, it seemed, was far from being just a scorched husk.

Mariner was able to sense the remains of an atmosphere consisting primarily of helium, as well as a magnetic field and a large iron-rich core, opening a mystery that would remain unexplored for another 30 years. As it flew past Mercury for the last time on 16 March 1975, the transmitters were switched off and its contact with Earth silenced. Mission completed, Mariner 10 began a lonely orbit of the Sun that, as far as we know, continues to this day.

Opposite: Mariner 10, launched on 4 November 1973 from Cape Canaveral, was the first unmanned spacecraft to fly past Mercury, managing to map half of the planet's surface in the process through over 2,800 photos, and giving us a unique insight into its history and makeup.

Above: The first glimpse of Mercury from Messsenger, nearly 40 years after Mariner 10's historic mission.

'We have Mercury in our sights.'
MDIS Instrument team, 10.30 am EST, 9 January 2008.

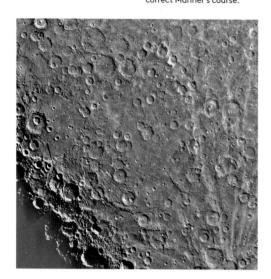

Below: Mercury's elliptical path around the Sun shifts slightly with each orbit, such that its closest point to the Sun moves forward with each pass. This discovery could not be verified with Newtonian physics and it took Albert Einstein's theory of relativity to finally explain it.

MERCURY'S UNIQUE ELLIPTICAL
ORBIT OF THE SUN

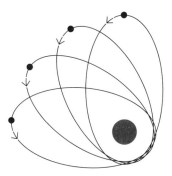

'With the beginning today of the primary science phase of the mission, we will be making nearly continuous observations that will allow us to gain the first global perspective on the innermost planet.'
Sean Solomon, Messenger mission

At first sight many things about Mercury simply don't make sense. During its 88-day orbit around the Sun it travels in a lopsided, elliptical orbit, which means it can be as far away from the Sun as 70 million kilometres but occasionally as close as 46 million kilometres. This is by far the most irregular of orbits of all the planets, but it is not the end of Mercury's oddity. Temperatures at midday can rise to 430 degrees Celsius on the surface, but at night, because it's a small planet and it has no atmosphere, temperatures fall to minus 170 degrees, giving it the greatest temperature swing of any known body in the Solar System. Its rotation is also unusual, gravitationally locked to the Sun in what is known as a 3:2 spin orbital resonance. This means the planet spins precisely three times on its axis for every two orbits, which in turn means that its day is twice as long as its year. In effect, you could be travelling over its surface at walking pace and keep the Sun at the same point in the sky as you strolled through eternal twilight.

As planetary scientist Nancy Chabot explains, 'A day on Mercury is not like a day on Earth. It has a very unusual orbit ... It has to go around the Sun twice to have one complete solar day on the planet, where the Sun goes from directly overhead to directly overhead and this actually takes 176 Earth days.' Because of the planet's orbit, there are places on the Mercurian surface where a hypothetical observer would be able to see the (two and a half times larger in the sky) Sun appear to rise and set twice during one Mercurian day. It rises, then arcs across the sky, stops, moves back towards the rising horizon, stops again, and finally restarts its journey towards the setting horizon.

Most of Mercury's anomalies can be explained by the orbital mechanics of its journey around the Sun, except, that is, for the odd elliptical orbit that takes it on such an oval-shaped, elongated course. This irregularity has puzzled astronomers for centuries and hints at an ancient planet that was very different from the Mercury we see today.

> *5 ... 4 ... 3 Main engines start 2 ... 1 ... and zero and lift off of Messenger on NASA mission to Mercury ... a planetary enigma in our inner solar system*

To truly begin to understand Mercury's history we had to wait nearly 40 years before we could return to her. On 18 March 2011, NASA's Messenger spacecraft became the first to enter Mercury's orbit, and over the next four years it succeeded in not only photographing 100 per cent of the planet's surface, but also collecting extensive data on its geology.

But before any of this could happen, Messenger had to take perhaps the most circuitous route in the history of our exploration of the Solar System. Just passing close to Mercury to take a few snaps, as Mariner 10 did, is hard enough, but actually entering into its orbit was thought to be either too difficult to achieve or too costly to execute. As cosmochemist on the Messenger mission Larry Nittler explained, 'There are two major challenges to getting a spacecraft into orbit around Mercury: gravity and money. When you go from Earth to Mercury, you're falling into the gravitational well of the Sun, which makes you accelerate faster and faster as you get closer. And, if you were to go straight from Earth to Mercury, this means that you would basically just zip right by the planet, or you would need to bring an incredible amount of fuel to put the brakes on, more than you could actually afford.'

Above: At the end of a long, circuitous route, Messenger finally enters Mercury's orbit on 18 March 2011, the first spacecraft to do so.

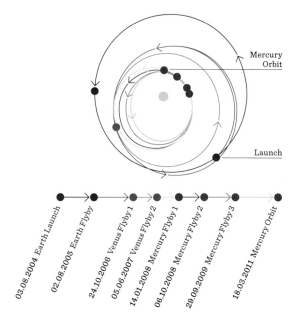

Mercury Orbit

Launch

03.08.2004 Earth Launch
02.08.2005 Earth Flyby
24.10.2006 Venus Flyby 1
05.06.2007 Venus Flyby 2
14.01.2008 Mercury Flyby 1
06.10.2008 Mercury Flyby 2
29.09.2009 Mercury Flyby 3
18.03.2011 Mercury Orbit

MESSENGER ENTERS MERCURY'S ORBIT

A number of missions never made it further than pencil and paper, while others floundered and failed at the proposal stage. It was only when Chen-wan Yen, a NASA engineer from the Jet Propulsion Laboratory (JPL), provided a trajectory that could not only get a craft into orbit but could do it at an estimated bargain-bucket cost of 280 million dollars, that the Messenger mission could really begin to take flight.

Taking off from Cape Canaveral on 3 August 2004, Messenger began a six-year, seven-month, 16-day journey to Mercury that would take it on a 7.9-billion-kilometre trajectory before it entered into orbit around the smallest of all the planets. To arrive at Mercury with the right speed and on the right course would require a complex route that would entail a number of gravity-assist manoeuvres around the Earth, Venus and Mercury itself to reduce the speed of the craft relative to Mercury. So, combined with the brief firing of its large rocket engine to finally insert it into orbit, this mission profile allowed Messenger to complete its voyage without the need to carry the vast reserves of fuel required to slow its passage through the firing of rockets. This design made the craft lighter and cheaper, but ultimately much slower. Almost seven years was a long time to wait for the team patiently charting its progress across the stars. Larry Nittler described Messenger's course as 'sneaking up on [Mercury] by taking a seven-year journey, flying around the Sun many times, doing multiple flybys around Mercury and Venus, and each time

Diagram top: Messenger's six-year, seven-month, 16-day journey to Mercury took it on a complex route involving several gravity assist manoeuvres before it entered the planet's orbit.

Diagram bottom: The highly elliptical path taken by Messenger to finally enter Mercury's orbit at 00.45 UTC on 18 March 2011.

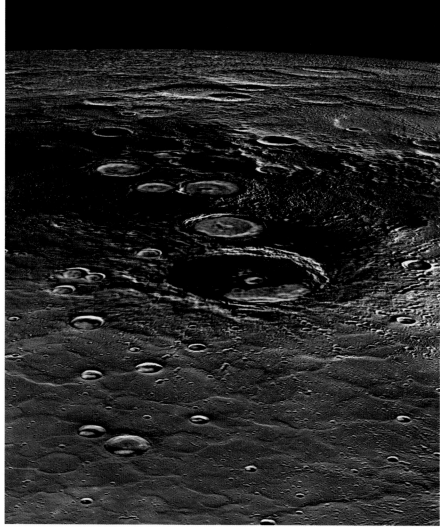

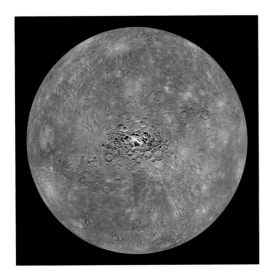

transferring some of [the] craft's speed and energy to the planet, so it could slow down, so that when we finally got to Mercury after seven years, we were able to fire our engine just a little bit, to slow down [even more] and get captured by the weak gravitational field of the planet'.

Appropriately, when Messenger finally entered into Mercury's orbit at 00.45 UTC on 18 March 2011, the path it settled into was highly elliptical. This orbit took it on a 12-hourly cycle from 200 kilometres above the planet's surface to 10,000 kilometres away from it. It may seem like an odd orbit for a craft with the singular aim of getting as close to Mercury as possible, but this was an essential part of the design of the mission, vital to protect Messenger from the fierce heat radiated by the scorching hot surface of Mercury. The sunlight reflected from the surface is so powerful it would have literally melted the solder holding the spacecraft together if it wasn't given time to cool down between its closest approaches to the planet.

Protected by an enormous ceramic solar shield and its eccentric orbit, Messenger could begin its work. For two years the spacecraft mapped pretty much every bit of the surface of Mercury, and the images beamed back to Earth revealed a planet that's been in the firing line for billions of years. Too small to hold on to an atmosphere that might protect it from meteorites, and lacking any processes to recycle old terrain, Mercury's ancient surface is the most cratered place in the Solar System.

Cosmochemist Larry Nittler explains the reason behind Messenger's elliptical orbit
'The way we addressed the problem of heat from the planet was to be in an extremely elliptical orbit, where we flew in very close over the North Pole, and took observations close to there, but then flew very far over the South Pole, like 10,000 kilometres. And so a couple of times a day we'd zoom in over the North Pole, get our data close, but the instruments would heat up, so then we'd fly and get different data farther out from the planet while we cooled, and in this way – heat up, cool down – we kept everything below the danger temperatures where instruments could be damaged.'

Opposite: Messenger's mission was a deeper exploration of the cratered landscape and geology of Mercury. One major discovery from its imaging work was evidence of water ice in its polar craters.

Top: Messenger took images of Mercury's south pole on several orbits, allowing scientists to monitor the region through changing illumination.

Above: This computer photomosaic of Mercury's southern hemisphere was created from images taken by Mariner 10 on its flyby of Mercury, giving scientists a tantalising glimpse of this elusive planet.

MAPPING MERCURY

'Scars are just another kind of memory.'
M.L. Stedman

The Mariner 10 mission had enabled scientists to see about half of the planet, so the first full view of the terrain of Mercury came from the flybys of Messenger. As planetary scientist Nancy Chabot explains, 'Before Messenger, we had only seen 45 per cent of the planet and we saw some stuff during the flybys before we went into orbit, but after orbiting the planet we have now mapped 100 per cent of the planet and seen nearly everywhere. There are some permanently shadowed regions which are still mysterious ... but after mapping the full planet, we have a good idea of what the surface looks like and craters are absolutely a dominant land form. This planet has been sitting there for billions of years and been hit over and over, and it hasn't had a lot of processes to destroy those craters.'

Amongst the thousands upon thousands of craters on Mercury, the largest by far is Caloris Planitia, a lowland basin 1,525 kilometres in diameter that is thought to have formed in the early years of the Solar System, around 3.9 billion years ago. It was first spotted as Mariner 10 sped past in 1974, but due to the trajectory and timing of the craft only half of it was lit, so the full character of this crater remained

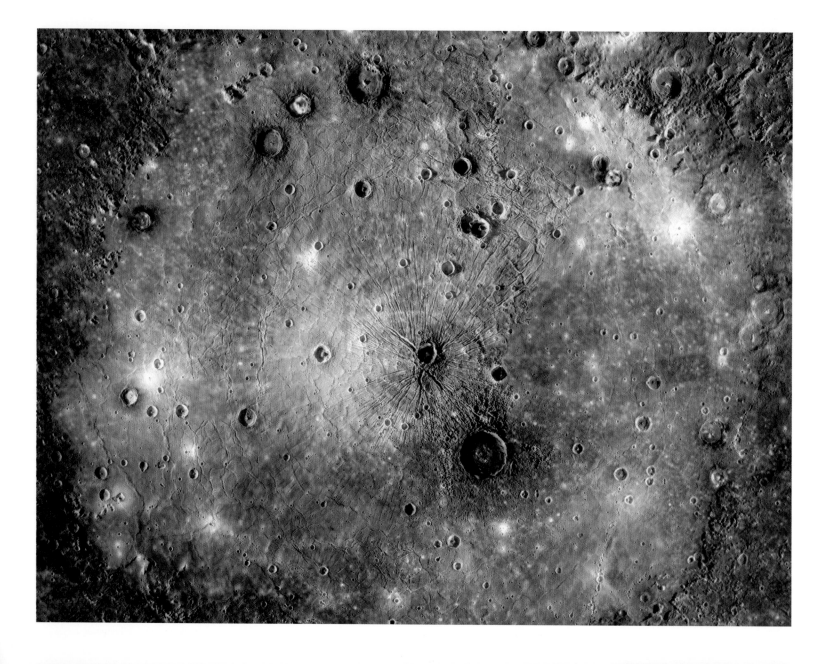

a mystery for another 30 years until Messenger could photograph it in all of its glory. Taking one of its very first photos, Messenger revealed Caloris to be bigger than had been previously estimated, encircled by a range of mountains rising 2 kilometres from the Mercurian surface, whose peaks create a 1,000-kilometre boundary around the lava plains within.

On the other side of the mountains, the vast amount of material that was lifted from the planet's surface at the moment of impact formed a series of concentric rings around the basin, stretching over 1,000 kilometres from its edge. The collision that created Caloris hit Mercury with such force that it also had more global consequences. Messenger photographed in great detail an area named (in the not particularly scientific vernacular) 'the weird terrain', a region at the planet's diametrically opposite point, the antipode, to Caloris. This area of strange geological formations distinct from the rest of the surrounding terrain was likely created by the seismic shockwave of the Caloris impact reverberating through the whole of the planet.

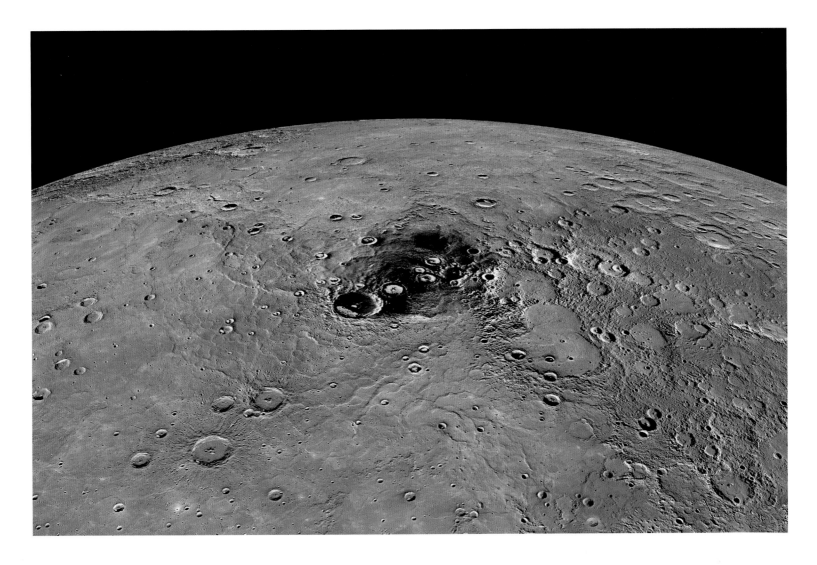

Opposite: This colour mosaic of Mercury's Caloris basin was created using images taken by Messenger in 2014.

Above top: Messenger photographed Mercury's geology in great detail, capturing this crater within the vast Caloris basin.

Above: In this 3D view of Mercury's north polar region, the areas marked in yellow show evidence of water ice.

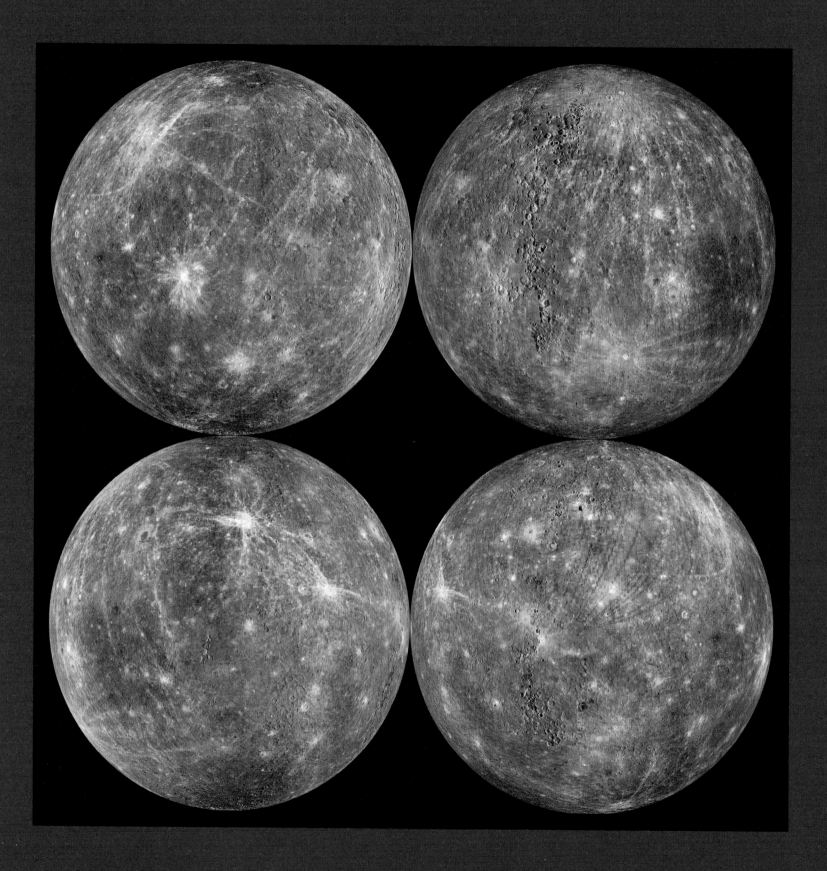

'We couldn't quite believe it, in fact we thought the data was wrong … we spent over two months looking at and double-checking the information but it was correct, Messenger had found a high level of volatiles such as sulphur, sodium and potassium on the surface.'
Nancy Chabot, planetary scientist, Messenger mission

Right up until the end of its mission in 2015, Messenger continued to uncover many of Mercury's secrets, including a few very particular surprises. Using a combination of photography, spectroscopy and laser topography, Messenger revealed tantalising evidence that even this close to the Sun, water ice can exist on the surface of a planet. Even though the Sun blasts much of Mercury's surface, the tilt of its rotational axis is almost zero, so there are craters and features around the planet's poles that never see direct sunlight. Combined with the lack of atmosphere, these regions are forever exposed to the freezing temperatures of space, and it's in this environment that Messenger was able to record the clear signature of water ice. Here, in the eternal night of a polar crater, it's cold enough for ice to survive for millions of years, just metres away from the savage ferocity of the Sun's light.

However, Messenger's most startling discovery was still to come. The mission objectives had been developed to explore the deep history of Mercury and provide data to test against our theories of the formation and early life of the planet. Messenger was equipped with a collection of spectrometers designed to analyse the composition of Mercury at different depths. The Messenger team had worked on a detailed set of predictions outlining the chemistry of the planet, but as the spacecraft began to sniff at the Mercurian surface it soon became clear that our assumptions had not been quite right.

As the gamma-ray and X-ray spectrometers analysed the elements on Mercury's surface they began to measure the unexpected characteristic signature of a number of elements such as phosphorus, potassium and sulphur at much higher levels than they were expecting. Up to this point, the working hypothesis had been that during the formation of Mercury (and all the rocky planets), as the rock condensed and combined to form the planet, the heavier elements like iron would sink towards the centre, forming the bulk of the core, while the lighter elements, such as phosphorus and sulphur, would remain near the surface. These more volatile elements would then be expected to be stripped away from the surface, particularly on a planet like Mercury, which is so close to the Sun. And yet the Messenger data confirmed high levels of potassium, and sulphur was detected at ten times the abundance of the element on Earth or the Moon. Both are volatile elements, easily vaporised, and when this close to the Sun, they simply should not have survived the planet's birth.

On top of that, the Messenger data confirmed what we had long suspected about the structure of Mercury, that it is the densest of all the planets, with a massive iron core making up 75 per cent of the planet's radius compared to just over 50 per cent here on Earth. The core creates a strange lopsided magnetic field, indicating that the internal dynamics of the planet are different to anything we have seen before.

All of this adds up to making Mercury something of a mystery, as nothing quite accords. The eccentric orbit, the abundance of volatile elements on the surface and the oversized iron core all point to the planet having a history far more complex than was first imagined, and the best explanation we have to make sense of the Messenger data is that Mercury was not born in its current sun-scorched position. It has long been supposedly known that the orbits of the planets are eternal, stable loops that sustain the structure of the Solar System in an endless rhythm, but everything we are learning now suggests that this is far from the whole story.

Opposite: The Mercury Atmosphere and Surface Composition Spectrometer (MASCS) instrument and the Mercury Dual Imaging System (MDIS) aboard Messenger enabled scientists to create these images, which use colours to map out the mineral, chemical and physical makeup of Mercury.

Above: Radio tracking data sent by Messenger has enabled scientists to create maps of the gravity field of Mercury. In this image, Mercury's gravity anomalies are depicted in colours: red indicates mass concentrations around the Caloris basin (centre) and the Sobkou region (right).

MEASURING MERCURY

Messenger was equipped with seven scientific instruments to collect data, including the Mercury Dual Imaging System (MDIS), Gamma-Ray and Neutron Spectrometer (GRNS), X-Ray Spectrometer (XRS), Magnetometer (MAG), Mercury Laser Altimeter (MLA), Mercury Atmospheric and Surface Composition Spectrometer (MASCS) and Energetic Particle and Plasma Spectrometer (EPPS). All these instruments communicated with the spacecraft through Data Processing Units (DPUs) and had to be mounted on the spacecraft with a view of Mercury but without interference from the Sun. They were designed to withstand the extreme temperatures the craft would encounter.

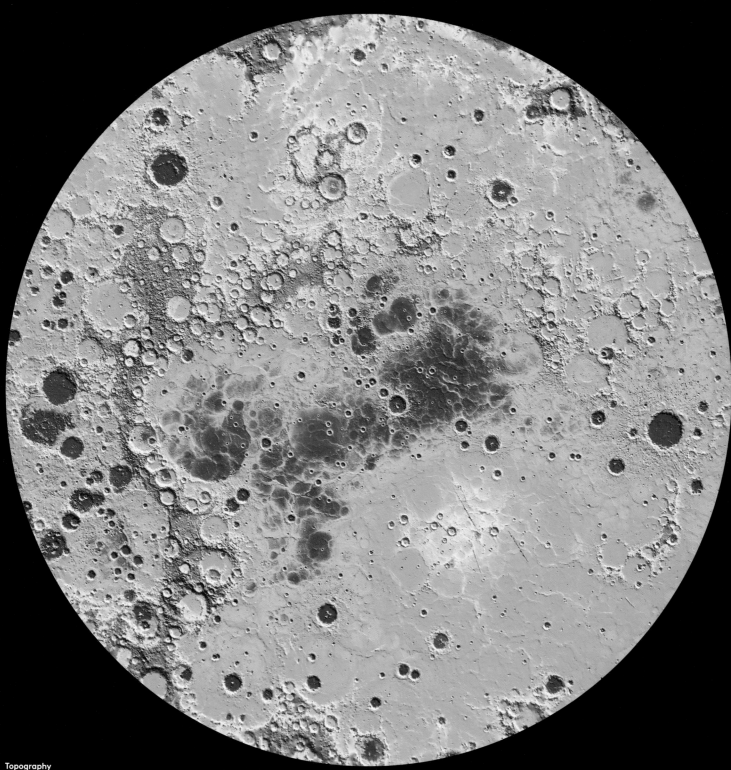

Topography

Messenger's MLA equipment was able to measure the difference in elevation across the northern hemisphere of Mercury, revealing it to be 10 kilometres between the lowest and highest regions.

low   high

Temperature
Messenger recorded expected information about the temperature of the planet, that the craters which were sunlit reached high temperatures, reflected in the red colouration of these images.

-223°C   >130°C

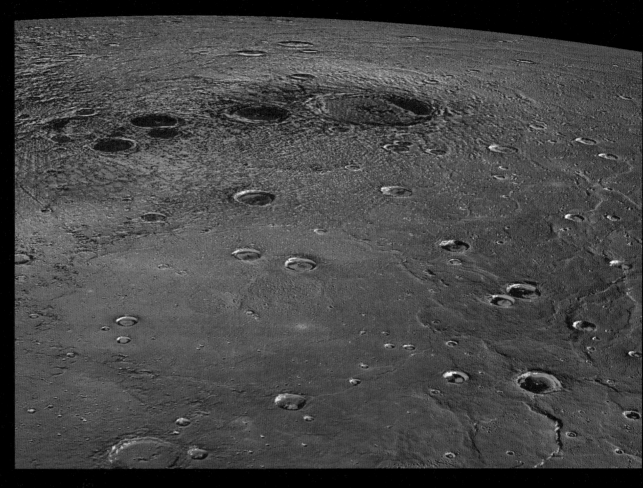

Geology
In this enhanced colour mosaic, the smooth volcanic plains of the Caloris basin are coloured yellow, with the craters picked out in blue.

crater   plain

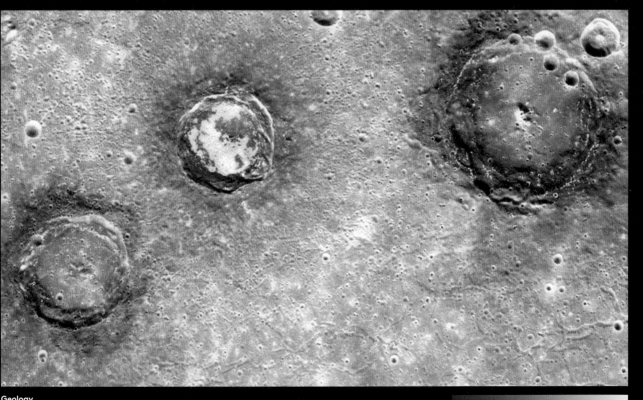

A SECRET HISTORY

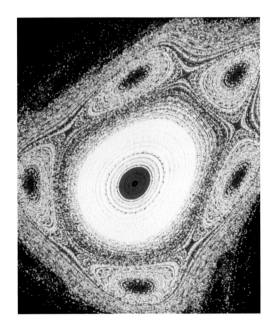

Mercury, like all four of the rocky siblings, was formed of molten rock. A few million years later, as the young planet began to cool, its crust solidified and its journey around the Sun transformed from being part of a swirling cloud into a clearly defined passage, an orbit. The path the infant Mercury travelled, however, was most probably far removed from the course it now holds. The young Mercury was born not as the closest planet to the Sun but at a much greater distance, far beyond the orbit of Venus, beyond Earth, perhaps even beyond Mars. This was a planet that came into being in the mildest region of the Solar System. It was far enough away from the Sun to allow volatile elements like sulphur, potassium and phosphorus to be folded into its first rocks without being vaporised away by the heat of the Sun, but maybe near enough for its surface to be warmed, perhaps even just the right amount for liquid water to settle on its surface. This may well have been a planet big enough to hold an atmosphere, a watery world upon which all the ingredients of life could well have existed. Mercury, it seems, really did have its own moment in the sun, but these hopeful beginnings were not to last.

Today it's hard to imagine the planets in any orbit other than our night sky. They feel eternal, permanent, and so it's natural to think of the Solar System as a piece of celestial clockwork, a mechanism running with perpetual and unchanging precision, marking out the passage of time. In time frames that we can comprehend – days, weeks, months and years – the motion and trajectory of the planets is just that: clockwork. We use these markers to plot out the 24 hours of a day, 365 days of a year, and the lunar cycle is, of course, intimately linked to our months. Beyond that, Newton's laws of universal gravitation first described in 1687 allow us to this day to plot out the trajectories of all the heavenly bodies far into the future and back into the distant past. This predictability of motion is what allows us to plot great astronomical events, such as eclipses and transits, far into the future. It's why, for example, we can predict that on 14 September 2099 the Sun, Moon and Earth will be in precise alignment to create the final total solar eclipse of the twenty-first century across North America.

But 100 years ahead or behind us is nothing more than a proverbial blink in terms of the life of the Solar System, and over longer durations the clockwork becomes a lot less reliable. If there was only one planet orbiting one star – for example, if Mercury was the orphan child of the Solar System – we would be able to calculate precisely the gravitational force between Mercury and the Sun, and to plot Mercury's orbit around the Sun with essentially infinite precision. But add one more planet into our rather vacant imaginary solar system – let's say we make it Jupiter – so there is now a gravitational force between all three objects – the Sun, Mercury and Jupiter – and it's no longer possible to calculate exactly where they're all going to be in the future or where they were at some point in the past.

'One possible theory is that Mercury didn't form where it is today, but much closer to the other planets, maybe even outside of Venus, or Earth, or somewhere in between. Then because of interactions with Jupiter, Earth, Venus, and so on, it got put into a chaotic path that pushed it farther into the Sun.'
Larry Nittler, cosmochemist, Messenger mission

Opposite: As predictable as the sunrise, Mercury keeps its place in our solar system, visible in the glow of dawn over Haleakala National Park, Hawaii.

Above: Chaos theory is used to predict the development of large-scale events from a given starting point, as shown in this Henon mapping of a chaotic system.

<u>The incredible shrinking planet</u>
The surface of Mercury is made up of just one
continental plate covering the entire planet. Over the
billions of years since its formation at the birth of the
Solar System, the planet has slowly cooled, a process
all planets undergo if they lack an internal source of
heat renewal. As the liquid iron core solidifies, it cools,
and the overall volume of Mercury shrinks.

When NASA's Mariner 10 mission circled the planet
in the 1970s, it captured images of surface features
created by the shrinkage. The contracting planet pushed
the crust up and over itself, forming scarps that can
extend miles below the planet's surface. At the same
time, the shrinking surface caused the crust to wrinkle
up on itself, forming so-called 'wrinkle ridges'.

The scarps and wrinkle ridges identified by Mariner
10 allowed scientists to estimate that the planet had
lost approximately 1 to 2 kilometres in global radius,
a finding that contrasted with their understanding
of the heat loss the planet suffered over time.

When there are more than two objects in play at any one time you have what
physicists call a chaotic system. It means the planets can push and pull one another,
moving entire orbits in ways we simply cannot predict. So the further we look back
in time, the less certain we are of the position of any of the planets. Our mathematics
fails, so instead we have to rely on circumstantial evidence to piece together a
picture of the past. In the case of Mercury, it's the evidence from Messenger detailing
the levels of volatile elements like potassium and sulphur that enable us to begin
to understand the early life of the planet and infer that Mercury must have begun
life further out in the Solar System than it finds itself today. So what happened next?
How did a planet that began its life in the sweet spot of the Solar System end up in
the scorched interior?

The answer lies in the other clue Messenger confirmed for us – Mercury's
massive iron core. Relative to its size, Mercury has the most massive core of any of
the rocky planets: 75 per cent of its diameter and almost half of its mass is molten
iron, compared to around just a fifth of the mass of the Earth. We've suspected
the oddity of Mercury's composition for well over 150 years, and that's because of
some brilliant deduction by a German astronomer called Johann Franz Encke,
who determined the mass of Mercury by measuring the gravitational effect it had
on a passing comet, a comet that we now call, unsurprisingly, Comet Encke. With
an approximation of the planet's mass we are able to calculate the density of the
planet, and with that calculation approximate its composition.

So we've known for some time that Mercury is odd, but only with the arrival of
Messenger did we begin to reveal just how odd the smallest planet actually is. By
accurately measuring Mercury's magnetic field we've been able to confirm that far
from being a geologically dead planet, Mercury has a dynamic magnetic field driven
by an internal force, indicating that the core is at least partially liquid. This goes
against the conventional thinking of planetary dynamics because we would expect
a planet as small as Mercury to have lost its internal heat long ago. Just as Mars lost
its heat because of its size (a story we will come to in the next chapter), we would
have expected the core of Mercury to have cooled and solidified.

But Messenger's data proved otherwise. By combining precise measurements of
Mercury's gravity field with the extraordinary mapping of its surface, Messenger

ANATOMY OF MERCURY

2440 km

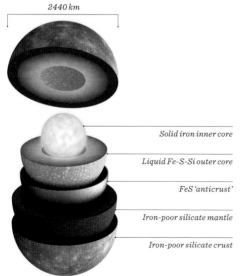

Solid iron inner core

Liquid Fe-S-Si outer core

FeS 'anticrust'

Iron-poor silicate mantle

Iron-poor silicate crust

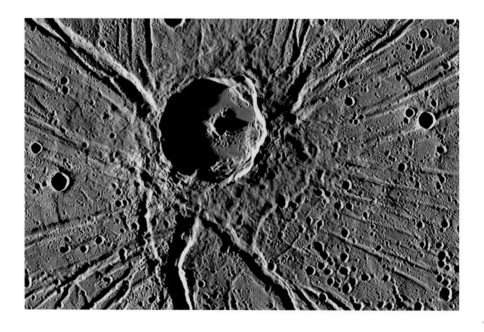

MESSENGER IMPACT SITE

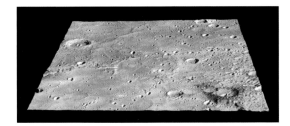

ESTIMATE AS OF 24 HOURS BEFORE MERCURY IMPACT

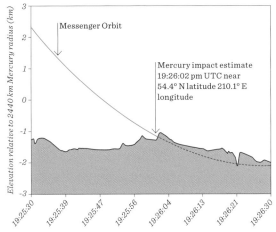

Time on 30 April 2015 (spacecraft UTC)

found that Mercury's structure is unique in the Solar System. It appears to have a solid silicate crust and mantle above a solid layer of iron sulphide, which surrounds a deeper liquid core layer, possibly with a solid inner core at the centre of the planet. This challenges all the theories about its formation.

Four and a half billion years ago, we know that the inner Solar System was in turmoil. In the middle of it all, we think that the newly born Mercury found itself orbiting far out from today's intimate proximity with the Sun, surrounded by rocky debris and scores of planetary embryos all jockeying for position. The young Solar System was still a place where planets could live or die. But it wasn't just the rocky planets that found themselves disturbed; Jupiter, the largest and oldest of all the planets, was on the move, and when a planet of that size shifts its position there are almost always casualties. We'll come back to the story of Jupiter's grand tack and the havoc it spread throughout the Solar System in Chapter 3, but for now all we need to know is that the evidence suggests that the juvenile Mercury was kicked by the gravitational force of Jupiter on an inward trajectory, finding itself flung in towards the Sun and into the path of danger. In the crowded orbits of the early Solar System such a change of course was fraught with danger, and all of the evidence indicated that this was the most violent and defining of turns in Mercury's history. As the planet swerved inwards it collided with another embryonic world and shattered.

Today we see the evidence of this ferocious collision in the strange structure of this tiny planet. A giant core has been left behind, the exposed interior of a planet that had much of its outer layer, its crust mantle, stripped away and lost to space in the aftermath of the collision. This collision not only transformed the physical characteristics of the planet but also knocked Mercury further inwards on a lopsided trajectory that we see reflected in the most elliptical orbit of all the planets. Although we cannot be certain of these events, it's a brilliant piece of scientific deduction to use the evidence we have to create a plausible scenario of events that happened unimaginably long ago. Events that drove the first rock from the Sun from a position full of potential to a place much too close to the Sun to support any form of life; an opportunity lost. After four years of observation and its investigation of Mercury's ancient past, Messenger finally ran out of fuel on 30 April 2015, and added yet another crater to this tortured world that once held such promise.

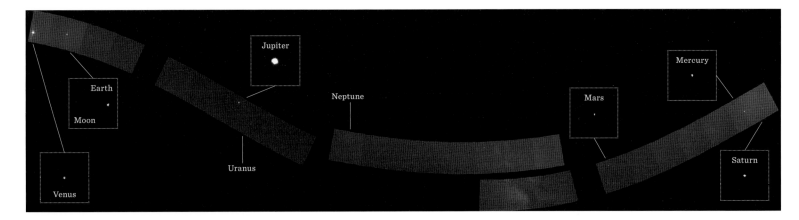

Opposite: Messenger captured this image of Apollodorus crater, near the Caloris basin; the radiating troughs led scientists to give it the nickname 'the spider'.

Above: In 1990, Voyager 1 captured a series of images from which we could create a portrait of the Solar System, giving a clear location for Mercury and its distance from Earth.

Mariner 10 took the first close-up
images of Mercury (above) and
Venus (opposite) in the 1970s, which
allowed scientists to compare the
planets' atmospheres.

PLANET OF MYSTERY

'I can find no reason ... for denying that she may be considered the abode of creatures as far advanced in the scale of creation as any which exist upon the Earth.'
Richard Proctor, English astronomer, 1870

Above: A hazy and cloud-shrouded Venus, photographed by Pioneer Venus Orbiter.

Shrouded in an unbroken blanket of cloud, the next rock from the Sun tells a very different story. Over 50 million kilometres beyond Mercury lies a world that at first sight has the potential to be far more Earth-like than her scorched inner companion.

Venus is perhaps the most mysterious of all the planets, lying on the inner edge of the so-called 'habitable zone'; this is a planet that holds its secrets close. For centuries it has teased us with its brightness in the early morning and early evening sky. It's so bright because it's a large planet about the same size as the Earth, it's not too far away from us either, and the clouds that shroud it are highly reflective, reflecting three-quarters of the light that hits them. That's the frustrating but tantalising thing about Venus, because even when you look at it through a big telescope, it is featureless; you never see the surface, which means that until the 1950s scientists could only speculate about what lay beneath.

In the late nineteenth and early twentieth centuries many thought that beneath her clouds Venus was hiding a mirror world to Earth; if not home to complex, sentient life, then certainly hosting basic life forms. Faced with that impenetrable cloak, our collective imaginations fuelled the idea of a living, breathing world beneath the clouds, a shroud that meant for the first half of the twentieth century we lived convinced that we were far from alone in the Solar System.

Nobel Prize-winning chemist Svante Arrhenius was one of the most renowned scientists to fuel the mythology of what lurked behind Venus's cover. Like many of the scientists of his era, Arrhenius let his curiosity wander into many different realms, including astronomy, and he hypothesised at length about the Venusian environment. Assuming the clouds of Venus were composed of water, he wrote in his book *The Destinies of the Stars* that 'a very great part of the surface of Venus is no doubt covered with swamps', creating an environment not unlike the tropical rainforests found here on Earth.

Svante Arrhenius, Nobel Prize-winning chemist
'*Everything on Venus is dripping wet ... A very great part of the surface ... is no doubt covered with swamps corresponding to those on the Earth in which the coal deposits were formed ... The constantly uniform climatic conditions which exist everywhere result in an entire absence of adaptation to changing exterior conditions. Only low forms of life are therefore represented, mostly no doubt belonging to the vegetable kingdom; and the organisms are nearly of the same kind all over the planet.*'

Expanding on this picture, he suggested that the complete cloud cover of the planet created a uniformity totally unlike the extremes of weather that define different parts of the Earth. In Arrhenius's imagination this stable environment, with a consistently uniform climate all over the planet, meant that any life on Venus lived without the evolutionary pressures of changing environments that drive natural selection here on Earth, leaving Venus in an evolutionary limbo akin to the Carboniferous Period. Describing a world full of prehistoric swamps and dank forests, Arrhenius created the perfect canvas for science-fiction writers of the time to conjure up a menagerie of curious life forms lurking beneath the clouds.

Today Arrhenius is far less known for his fertile imaginings on the wildlife of Venus than he is for his work on the climate of Earth. In 1896 he was the first scientist to use basic principles of chemistry to demonstrate the impact that the atmosphere can have, in particular levels of carbon dioxide, on the surface temperature, a process that was called the Arrhenius effect but is now known as the greenhouse effect. An effect that would not only have profound consequences for our understanding of our impact on our own planet, but would also be vital in explaining the true nature of Venus beneath the clouds.

By the 1920s, as ground-based technology improved, we stopped painting the surface of Venus with our imaginations and started filling in the gaps with facts. The first spectroscopic analysis of the planet's atmosphere suggested that it wasn't water or oxygen that filled the clouds of Venus, so some thought this hinted at an arid, desert land beneath. Others speculated that formaldehyde filled the air, leading to the belief that Venus was not only a dead planet but a pickled one, too. But come the 1950s the true nature of Venus began to be revealed, as more accurate Earth-based observation suggested the presence of overwhelming levels of one defining gas in the Venusian atmosphere. This was not a planet shrouded in clouds of water and oxygen, nor pickled in formaldehyde, this was a planet engulfed in a blanket of carbon dioxide, and as Arrhenius had demonstrated on Earth, this almost certainly meant that whatever lay beneath the clouds, the heat would be beyond the limits of even the most resilient life forms on Earth. As the first spacecraft were being built to explore our sister world, it was becoming increasingly clear that visiting Venus would be far from easy and she would be far from welcoming.

In the early 1960s, the Soviet Union began a series of missions under the programme name Venera, which attempted to explore the atmosphere and surface of Venus directly for the first time. The initial launches of the Venera programme failed before they had even left Earth's orbit, but within a couple of years the programme began to slowly see some success.

Venera 1 was successfully launched on 12 February 1961. Designed as a flyby mission, it is thought to have passed within 100,000 kilometres of Venus, but a total telemetry failure on the craft meant that no data was returned to Earth. As far as we know, Venera 1 is still in an orbit around the Sun to this day.

Venera 3 attempted to go a step further and was designed to enter the Venusian atmosphere to take the first direct measurements. However, on crossing the atmospheric boundary the probe's systems failed and no data was returned as it plummeted towards the ground. All that was left for Venera 3 was the historic position as the first human-built object to crash into another planet's surface.

Left: Artwork showing the successful landing of the Venera 9 spacecraft, which in its 53 minutes on the surface of Venus returned the first ever images of the planet.

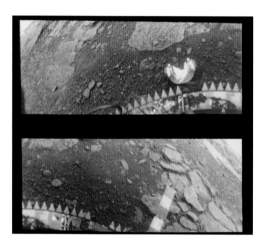

Despite multiple failures, the Soviets didn't give up and in October 1967 Venera 4 entered the atmosphere of Venus and sent back data supporting the Earth-based observations, revealing for the first time that the blanket of cloud surrounding Venus was made up of primarily carbon dioxide (90 to 95 per cent), 3 per cent nitrogen and just trace amounts of oxygen and water vapour. Venera 4 confirmed beyond all doubt that this was no second Earth: as it descended through the thick clouds, the temperature rose to 262 degrees Celsius, the atmospheric pressure increased to 22 standard atmospheres (2,200kpa) – and this was still 26 kilometres above the surface. As Venera 4 parachuted its way down to the surface it provided data back to Earth while confirming its own imminent death. This was a spacecraft that was not designed to survive the intense pressures and temperatures it was measuring, let alone the lack of the water landing it was designed for. The craft failed during the descent and was lost long before it reached the surface.

Gradually, through the following missions, the Soviet scientists began to overcome each and every challenge Venus put in front of them. Venera 7 was built to survive the most violent of landings, and even though its parachute failed, it made it to the surface intact in 1970 and was able to use its damaged antennae to transmit limited temperature data for 23 minutes before it expired.

Venera 9 not only made it to the surface and operated for 53 minutes in October 1975 but was also the first craft to successfully deploy its camera on the ground and transmit an image back to Earth. In the first-ever picture taken from the surface of another planet, the black and white fractured image revealed a rocky, desolate landscape with measurements confirming it to be a blistering 485 degrees Celsius, with an atmospheric pressure of 90 atm (standard atmosphere) crushing down.

By the time Venera 13 launched, on 30 October 1981, the ambition of the missions and the confidence in delivering data from the surface had been radically transformed. Venera 13 functioned for 127 minutes in recorded temperatures of 457 degrees Celsius and a pressure of 89 Earth atmospheres. The probe's cameras deployed, taking the first colour image from the surface of Venus, spring-loaded arms measured the compressibility of the soil, while a mechanical drill arm took a sample of the Venusian surface that was analysed by an onboard spectrometer. If that wasn't enough, onboard microphones were deployed to record the vicious winds that were assumed to be whipping the surface of Venus, the first-ever recording of the sound of another planet.

As the Venera missions came to a close in 1983, not even the smallest doubt remained of Venus's hostility. Far from the benign water world we had once imagined, the reality was that this was not a sister we recognised – in our search of the heavens for a place like home we'd found a toxic, fiery hellscape.

Venus is an enigmatic world – almost Earth-like in size, position and potential, and yet as far from paradise as it's possible to imagine. If Mercury's story is one of catastrophic orbital change and Earth's of balance and stability, the story of Venus is a tragedy; a tale of subtle, yet relentless decline. So why did it all go wrong for Venus? Why did a world born with such similarities to the Earth take such a different path? To answer that, we need to look beyond the tortured planet we see today and go back to a time when Venus was a young thriving planet.

Above: In March 1982, Venera 13 returned these photographs of the surface of Venus, with part of the spacecraft visible in the foreground.

Opposite: The Venera 1 display in the space (Kosmos) pavilion at the All-Russia Exhibition Centre, in Moscow, Russia.

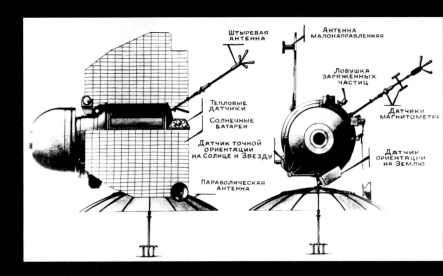

Through the following missions, the Soviet scientists began to overcome each and every challenge Venus put in front of them.

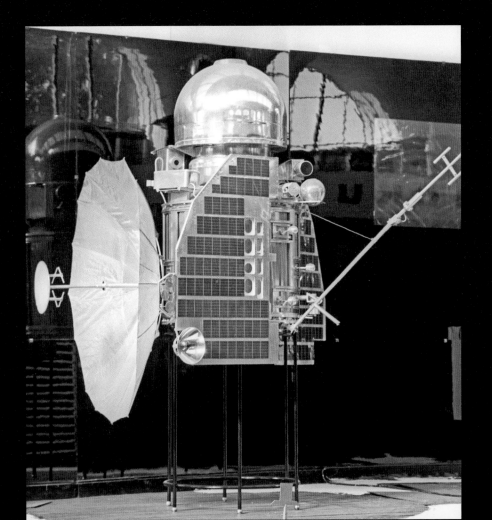

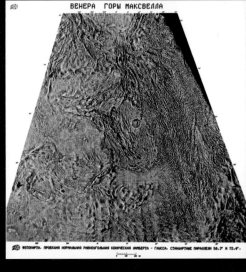

Above top: This radar image taken by Venera 15 and 16, offers a fascinating insight into the terrain of Venus, revealing the Maxwell Montes mountain range in the centre and the 100-km-wide Cleopatra crater.

Above bottom: Sediment and rocks visible on the landscape imaged by Venera 9.

THE BIRTH
OF VENUS

'Today Venus is incredibly hostile ... so hot, so dry, but what did it start out like, was it ever more Earth-like? We don't know for sure, so we want to make future spacecraft missions to nail down that early history.'
David Grinspoon, astrobiologist

Four billion years ago, Venus was a familiar world. A world created from the same dust as the Earth, born just about the same size and settled into an orbit that seemed just far enough away from the glare of the Sun to allow a precious process to begin to take hold. In almost every conceivable way, Venus's early life mirrored that of our own world. As its newly formed crust settled and cooled from the violent heat of its birth, an atmosphere began to grow around the young planet, fed by gases bubbling up from the molten rock below its surface, as well as captured from the clouds of gas and dust it swept through on its orbit around the Sun. Clinging to the young Venus, this thin layer of gas would have certainly contained nitrogen, oxygen and carbon dioxide, but most intriguing of all, we are certain it would have also contained large amounts of water vapour.

High in the Venusian atmosphere this water vapour eventually cooled enough to change state from vapour to liquid. And with that transformation, a process began, that perhaps for the first time on any of the planets would have seen the conditions become just right for droplets of liquid water to take shape and begin falling from the Venusian sky. These were the first rains of the Solar System, showering down onto the dry plains of Venus. Gradually these rains would have not just fallen but flooded the surface, rivers would have flowed and shallow oceans taken hold of large swathes of the planet's surface. Venus, perhaps before even the Earth, became a water world, a planet with skies full of clouds and a surface full of oceans, feeding the cycle of water around this young planet.

How can we be certain this blue version of Venus existed? Unlike Mars, where we can see the evidence of its watery past etched onto its surface, we have no such direct evidence of the presence of liquid water on the surface of Venus. The only physical evidence we have suggests that the planet's watery past comes from measurements taken by NASA's Pioneer Venus spacecraft back in 1978. One of its most surprising discoveries revealed an unexpected amount of deuterium (heavy water) in the atmosphere compared with hydrogen. This D/H ratio is far smaller on Venus than it is on Earth, and that's interesting because when the two planets formed the ratio would have almost certainly been the same. Because hydrogen is far more easily lost from an atmosphere than deuterium, this smaller ratio suggests that Venus has lost a lot more water than the Earth over its lifetime – the signature of a long-lost primordial ocean. As cosmochemist Larry Nittler explains:

> 'Scientists believe that Venus once had a lot of water in its oceans, but lost it over time, and perhaps in oceans as recently as a billion years ago. The reason we can tell this is from the isotopic composition of hydrogen measured in its atmosphere by spacecraft. Now, hydrogen has two flavours of isotopes, whereas most hydrogen atoms are just a single proton in the nucleus. Some, a small fraction, are what we call deuterium, that have a proton and a neutron, so they weigh twice as much as the regular hydrogen. What happens when you have evaporation of water from a planet, or the atmosphere, is that the water molecules that contain hydrogen are much lighter than the water molecules that contain deuterium, so they evaporate more easily, and can be lost more easily. So, over time, as you evaporate water, deuterium-bearing molecules stay behind relatively to the regular ones, and you build up a deuterium to hydrogen ratio. And by back-calculating from the measured ratio today, we can figure out how much water has been lost over the billions of years of evolution, and [on Venus] it's quite a lot.'

Left: Rising centrally in this computer-generated image is the volcano Maat Mons, surrounded by cascading lava. This three-dimensional image was created using data relayed by Venera 13 and 14.

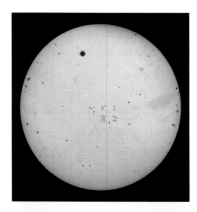

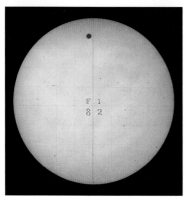

None of this is solid proof, but it does begin to point us in one direction, and with no further exploration of the surface we have had to rely on an accumulation of indirect evidence to begin to paint a more detailed picture of Venus's watery past.

As with almost all of our understanding of the planets, the evidence that built this picture has been accumulated through decades of exploration. Starting with the Venera missions' first touchdown on the planet to the Pioneer Venus orbiter, and to the more recent Magellan mission, which not only relayed extraordinary radar soundings of the surface of Venus but provided the first full topographical map of the planet collated over a period of four years in orbit.

Combining all of the data that has been accumulated over decades of exploration has allowed us to peer deep into the planet's past, using the same tools that enable us to model the future of climate change here on Earth to create climate models of Venus in the past, present and future. The results of this analysis, conducted most recently by a team from NASA's Goddard Institute for Space Studies (GISS), all point to the same conclusion – in the distant past Venus was a planet covered in shallow primordial oceans.

Some estimates suggest this water world was far from fleeting, a blue planet just like our own that could have been sustained for around 2 billion years and perhaps only disappeared some 700 million years ago. It's a tantalising thought that such a similar world to our own existed for so long with liquid water on its surface. We know life took hold quickly on our own blue planet, within half a billion years of the Earth being formed, so there seems good reason to suspect that if Venus really was as wet as the models predict, it too could have sprung into life. Exactly what went on in the long-lost rivers and oceans of Venus is yet to be discovered; hidden behind the

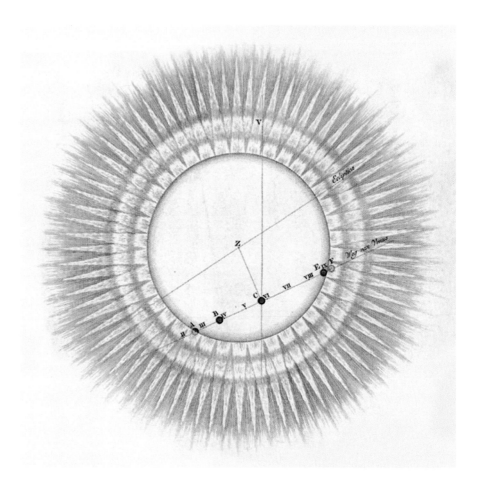

Above The transit of Venus – as the planet crosses the face of the Sun – was captured in photographic plates as early as 1882.

Right: Venus has fascinated scientists for centuries; this diagram was drawn by Nicholas Ypey in 1761, showing the transit of Venus that year.

'Oh most grateful spectacle, the realisation of so many ardent desires.'
Jeremiah Horrocks, seeing the transit of Venus in 1639

clouds, we have not yet been back to search for any signs that life ever took hold here. Our exploratory attentions have turned to Mars as a planet that not only has a fertile past but is also a possible target for human colonisation in the future. We know for certain that no life (at least no life we understand) could exist on Venus today, and perhaps even the evidence of any biology on that long-lost water world has long ago vanished under the oppressive heat, rampant volcanism and extreme pressures of the planet today. So where did all that water go? Understanding this requires an exploration of the differences between Earth and Venus, as well as the similarities.

Today Venus has the slowest rotation of any planet in the Solar System, taking 243 Earth days to complete one rotation on its axis. This period is known as the sidereal day, which is different to a solar day – the time it takes for the Sun to return to the same point in the sky. On Earth the sidereal day, at 23 hours, 56 minutes and 4.1 seconds, is very close to the solar day, which lasts pretty much exactly 24 hours. But on Venus the difference between these two periods is much greater. Even though the planet takes 243 days to rotate on its axis when combined with its orbit, a solar day on Venus lasts for 116.75 Earth days. It means every day on Venus lasts almost four months on Earth, and not only that, but Venus also rotates from east to west (one of only two planets to do so, along with Uranus). So across this toxic world a sunrise would last literally for days as it inches across the sky.

This slow progression of the Sun in the Venusian sky, due to the planet's creeping rotation, has raised many questions about how in the past the planet would have been heated and how the climate would have been affected by such a different rotation compared with the Earth's. Today the climate of Venus is what is known as isothermal – there is a constant temperature between the day and night sides and

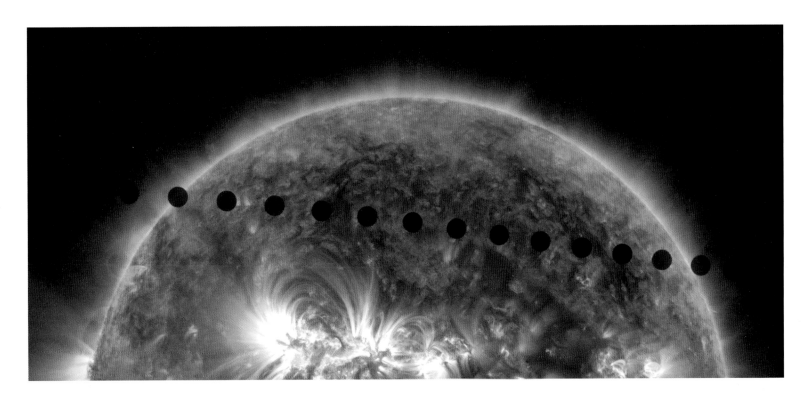

Top: Catching a glimpse of the transit of Venus is rare and has been important throughout the centuries. The next such sightings are predicted for 2117 and 2125. Recording each transit is vital research, which helps scientists to determine the scale of the Solar System.

Above: Composite image showing the transit of Venus in June 2012 (in black spots).

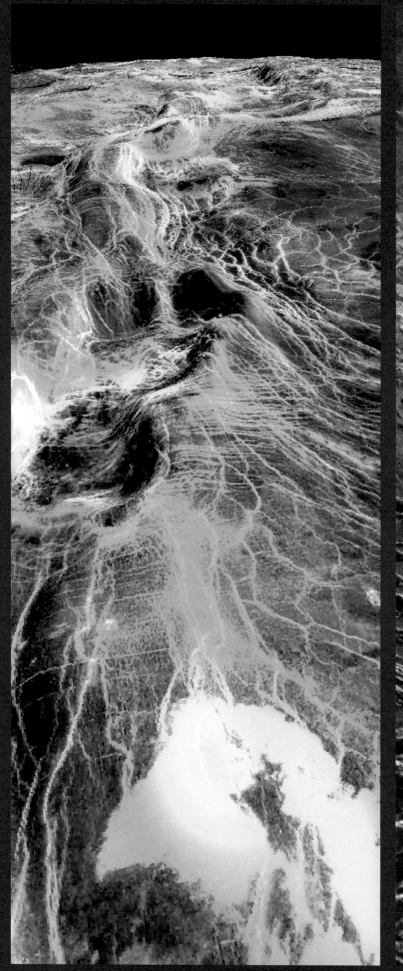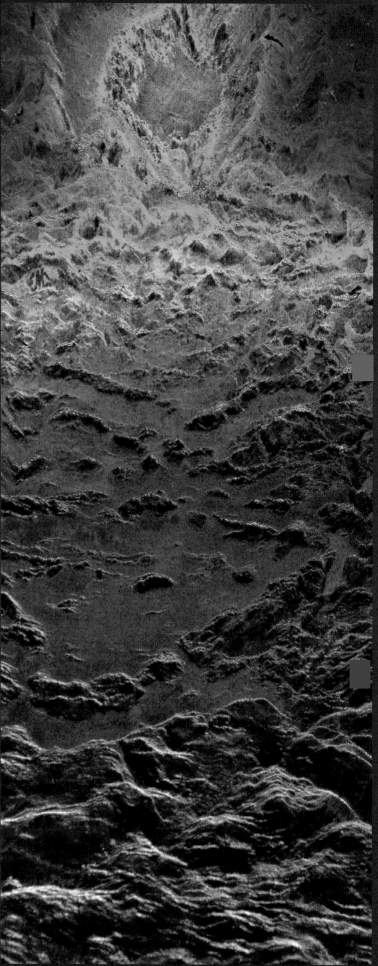

between the equator and the poles. This is because the thick atmosphere literally acts like a blanket, dissipating the heat of the Sun so that the only real variation in temperature on the Venusian surface occurs due to differences in altitude. In its past, however, this may have been very different – with a more Earth-like atmosphere, so the Sun would have been beating down on the planet's surface for days on end.

To make things even more complex, we know the spin of a planet is intimately linked to its climate and we've got strong evidence to suggest that how fast a planet spins is directly related to its chance of habitability. Until very recently it was assumed that the slow rotation of Venus must have been caused by the presence of a thick atmosphere early on in its history that in effect acted as a brake on the planet's spin. However, recent studies now suggest the planet could have had a thin atmosphere like that of modern Earth and still have ended up with its slow rotation.

Gradually, as we start to build a picture of ancient Venus we begin to see beyond the cloud cover of today through to an ancient planet with an Earth-like atmosphere, and a day lasting over 200 Earth days as the Sun beat down on the ocean-covered surface.

To make sense of the climate of this Earth-like Venus, the team at the Goddard Institute needed to make another tweak (or postulation, to be more precise) to the model. With the Sun hitting the one side of the surface for so much longer than on the Earth, the evaporation rate of the oceans would be far greater and potentially incompatible with the water world we suspect existed, but by simply adjusting the amount of dry land on the surface of Venus, especially in the Tropics, the effect is dramatic. With a higher percentage of land, the models suggest that even the slow rotation would not dry out the planet, and it could have held on to enough water to be ripe for supporting the emergence of life.

By combining all of this data, the GISS team have painted our most up-to-date picture of early Venus, and it's a beguiling image. Within the infant Solar System, it is a planet the size of Earth with a similar atmosphere to the one we see today. On Venus days lasted for months as the Sun arced slowly across the sky from west to east, rising and setting over a vast, shallow ocean.

Finally, the data from radar measurements taken by NASA's Magellan mission in the 1990s was used to paint the last brushstrokes of this long-lost world. Filling in the lowlands with water, the topography of this ancient world emerges with the highlands exposed as the Venusian continents. It all points to the possibility that Venus could have been the first habitable world in our Solar System. So what changed? To find out we need to look not just at the planet in isolation but also at the star around which it orbits.

'In the GISS model's simulation, Venus's slow spin exposes its dayside to the Sun for almost two months at a time. This warms the surface and produces rain that creates a thick layer of clouds, which acts like an umbrella to shield the surface from much of the solar heating. The result is mean climate temperatures that are actually a few degrees cooler than Earth's today.'
Anthony Del Genio, planetary scientist

Opposite and above: NASA's Magellan mission in the 1990s sent back images of Venus that enabled scientists to create a more detailed image of the landscapes of this long-lost world. They revealed a terrain of lowlands and highlands, dotted with active volcanos – a far cry from the ancient watery Venus imagined and depicted in some artworks.

GOODBYE TO LIFE

No planet lives out its life in isolation. Venus, like all the planets is part of a Solar System, a system that is driven more than anything else by the star at its centre. Today the Sun burns bright in our skies, bathing our planet in just enough starlight to keep the oceans from freezing, but not too much to boil them away. Earth lies in the sweet spot we call the Goldilocks zone, but as we have already seen in this chapter, nothing in the Solar System is forever and what we see today is not what we will see tomorrow nor what we would have seen yesterday.

VENUS

SEE YOU AT THE CLOUD 9 OBSERVATORY

VOTED *Best* PLACE IN THE SOLAR SYSTEM TO WATCH THE MERCURY TRANSIT

David Grinspoon, astrobiologist, on Venusian life:
'So when we say, as we often do, that Venus is completely uninhabitable, we should put a little asterisk next to that statement, because, we're talking about the surface environment. But actually if you go up from the surface about 50 kilometres you reach a zone that may be habitable on Venus; in the clouds the pressure and temperature is roughly what it is here on the surface of Earth. There are energy sources in terms of radiation and chemical energy, there are nutrients, there is even liquid water medium – although it's concentrated sulphuric acid in the clouds – but we now know of organisms on Earth that love concentrated sulphuric acid. So there's nothing to rule out life in the clouds of Venus and there are even some, I would say circumstantial, facts that suggest the possibility of a biosphere there. I wouldn't bet on life in the clouds of Venus, but I wouldn't rule it out until we've explored a little more carefully.'

THE GREENHOUSE EFFECT

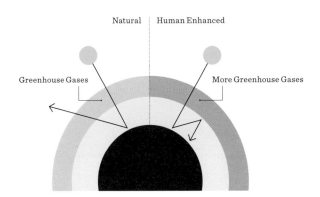

Natural | Human Enhanced

Greenhouse Gases | More Greenhouse Gases

As our Sun gets older, it's gradually burning hotter and hotter. This is because as it ages the process of nuclear fusion – the fusion of hydrogen into (mainly) helium – gradually leads to an increase in the amount of helium in its core. This rise in helium causes the Sun's core to contract, which in turn allows the whole star to shrink in on itself, creating an increased pressure that results in a rise in the rate of fusion, and so the energy output of the Sun goes up. If tomorrow the Sun is burning hotter than today, it of course makes sense that in the early days of the Solar System our Sun burned far less brightly. It's a life cycle that is common to all main-sequence stars, the category of star that includes our Sun, and as the most common type of star in the Universe we have been able to study this life cycle in intimate detail, allowing us to make immensely detailed predictions about the characteristics of our Sun in the past and in the future.

Winding back the clock, the current consensus amongst astronomers is that 4 billion years ago the faint young Sun was at least 30 per cent dimmer than it is today. This cooler Sun would have undoubtedly had a big impact on all of the terrestrial planets. Earth would have been much colder, and as it was receiving far less solar energy it remains something of a mystery as to why our planet wasn't frozen solid. Instead, at this time on Earth first life was just beginning, in the liquid water that we are pretty certain covered its surface.

At the same time, 3.5 to 4 billion years ago, the young Sun would have bathed Venus in a warmer glow. This ocean world found itself in its very own sweet spot, a world held in a delicate balance. With the Sun weakened and restrained, the Earth-like atmosphere of Venus could act as a gentle blanket, keeping the surface temperate and covered in an abundance of liquid water. But even with this additional solar energy we think Venus would have been cooler than the Earth is today; in fact we believe temperatures at that time would have been like a pleasant spring day here on Earth.

It wasn't to last. Slowly the young Sun grew brighter, its increased energy output causing temperatures to gradually rise, which in turn began to lift more and more water vapour into the air, thickening the atmosphere and sealing the planet's fate. Although the oceans of Venus may have persisted for billions of years, as the surface warmed and the atmosphere thickened, the destiny of this planet was already set, driven by an unstoppable process we have recently become very familiar with here on Earth.

The greenhouse effect is a process that has the power both to protect and to destroy a planet, but despite this power it actually boils down to some pretty simple physics. It's all about how sunlight – solar radiation – interacts with the constituent parts of an atmosphere. In the case of the Earth, as solar radiation hits our atmosphere some of it is reflected straight back out into space, some is absorbed by the atmosphere and clouds, but most of the sunlight (about 48 per cent) passes straight through the atmosphere and is absorbed by the Earth's surface, where it is heated up. The reason so much solar radiation makes it to the surface is because the gases in our atmosphere, like water vapour and carbon dioxide, are transparent to light in the visible spectrum. When you think about it, that's pretty obvious because there's a source of visible light in the sky, the Sun, and we can all see it! But it's a different story when that sunlight heats the surface of the Earth and re-radiates back out not as visible light but as the longer-wave infrared light – thermal radiation.

Opposite and diagram: Capturing the Sun's warmth is essential for life on a planet, but when too much heat is trapped it can have devastating consequences.

'Venus hasn't stopped heating up, and we believe that as the Sun continues to age, billions of years into the future, it's going to continue getting hotter. Eventually that means that Earth will go the way of Venus.'
David Grinspoon, astrobiologist

We can't see this light, but as it radiates back out from the Earth's surface, carbon dioxide and water vapour absorb the infrared, trapping that energy, and so the planet maintains a higher temperature that is intimately linked to the constituent parts of the atmosphere. The higher its level of gases like water vapour, carbon dioxide, methane and ozone, the greater the greenhouse effect and the bigger the uplift in temperature. Despite the very real threat that this now poses to the future of our planet, the greenhouse effect on its own is not necessarily a bad thing – the Earth would be at an average temperature of around minus 18 degrees Celsius without it – but as we are currently witnessing here on Earth, shift the balance of those gases and things can change very quickly.

At some point in Venus's past, the levels of water vapour lifted into the atmosphere by the warming sun pushed the greenhouse effect to become more intense. With less and less of the Sun's energy escaping, the ambient temperatures

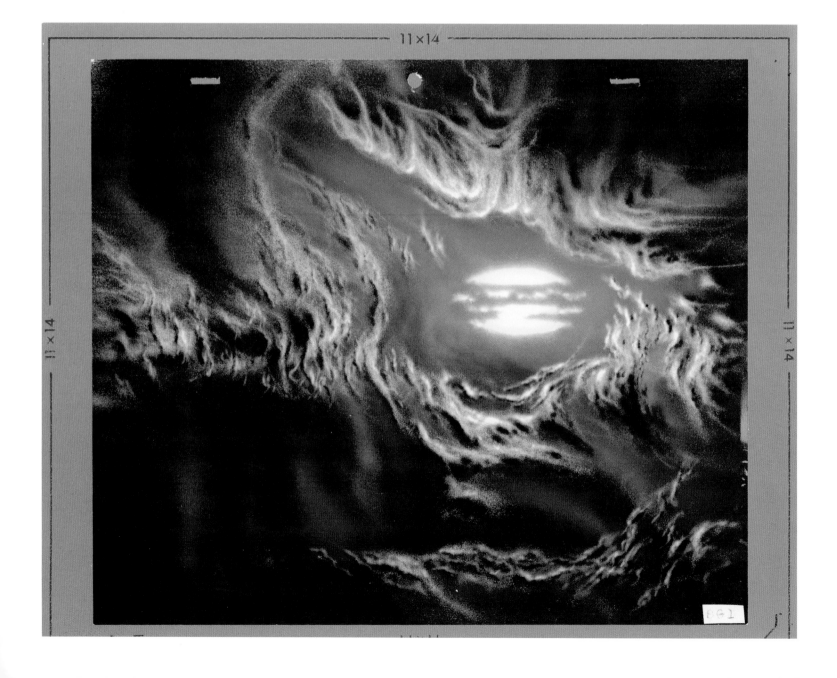

Opposite and below: In 1977, in the days before computer-generated imaging, NASA commissioned artist Rick Guidice to paint illustrations of the surface of Venus, based on images received from Pioneer probes.

began to rise exponentially until the day came when the last raindrops fell onto the surface of the planet, the heat evaporating the rains long before they could reach the ground. Venus had reached a tipping point: with the increasing temperatures feeding more and more water vapour into the atmosphere, a runaway greenhouse effect took hold, driving away the oceans. This led to the surface of the planet getting so hot that carbon trapped in rocks was released into the atmosphere, mixing with oxygen to form increasing amounts of another greenhouse gas – carbon dioxide. With no water left on the surface and no other means to remove it, carbon dioxide built up in the atmosphere, setting the planet on a course that would result in the scorched body that we see today.

And so Venus's moment in the Sun came to an end. Earthlings take note: when it comes to the greenhouse effect, there is a precariously thin line between keeping a planet warm and frying it.

THE END OF EARTH?

In a chaotic solar system, filled with planetary might-have-beens, Earth is a shining example of stability.

Above and opposite: Russia's Kamchatka Peninsula is one of Earth's most inhospitable areas; the volcanic landscape gives us an insight as to how planet Earth might appear when it becomes too hot for life.

Of the four rocky worlds, only one has managed to navigate through the instability and constant change of our Solar System over the last 4 billion years and maintain the characteristics needed to support life. Mercury lost its fight early as it was flung inwards towards the Sun, Venus flourished at first, before slowly coming to the boil, and Mars, the runt of the litter, became a frozen wasteland long ago. Only Earth, uniquely amongst the planets, has persisted with an adequate stability over the last 4 billion years to allow liquid water to remain on its surface and an atmosphere just thick enough to keep its climate calm – not too hot and not too cold. Events have rocked us and extremes of temperature have waxed and waned, but never outside of the parameters needed to harbour life. In a chaotic solar system, filled with planetary might-have-beens, Earth is a shining example of stability, and the evidence for this is to be found in every nook and cranny of the planet.

Today Earth is dominated by life; the land and seas are teeming with millions upon millions of species, with thousands of new life forms discovered each year. Somehow, even when disaster threatened, the Earth has remained a living world; while endless species have come and gone, life has always persisted. It's woven into the fabric of the planet – an integral part of every continent and every ocean. Life plays a crucial role in maintaining the balance of the atmosphere that keeps our planet temperate, but we know for certain it cannot last.

The Kamchatka Peninsula in Eastern Siberia is one of the most inhospitable places on Earth. A volcanic wasteland, peppered with thousands of hot springs, it's here that we find some of the toughest living things. Extremophiles survive here that are able to withstand temperatures and pH levels higher than any other land-based life forms we have ever discovered. Kamchatka is part of the Pacific ring of fire, and despite its remoteness, biologists have long been enticed here to explore its toxic, bubbling cauldrons for signs of life. Complex life, animals and plants struggle to survive in temperatures above 50 degrees Celsius, so searching for life here is all about searching for single-celled life forms, bacteria and archaea – ancient microorganisms – that are somehow able to endure in this hostile environment. Life forms like *Acidilobus aceticus*, an archaea that can be found in a hot spring where the water is so acidic it reaches a pH of 2, and where temperatures rise to 92 degrees Celsius. In other parts of the hydrothermal field, bacteria like *Desulfurella acetivorans* have been discovered, which happily live in pools that are touching 60 degrees Celsius, but it's these that are the real hotheads. In one of the biggest and hottest pools investigated by scientists, a large number of microbes have been found living in temperatures approaching 97 degrees – making it one of, if not the hottest environment ever studied for signs of life on land.

But to find the greatest hotheads on Planet Earth you need to look not on land but deep beneath the sea. In the furthest depths of the Atlantic, around the black smoker hydrothermal vents blurting out of the ocean floor, we've found strains of archaea that can survive temperatures of 122 degrees Celsius, and perhaps even higher.

These rare life forms live at the very edges of biology. Unique adaptations to their cellular chemistry enable the proteins and nucleic acids that create the structure of the microorganism to function, while the membranes that are protecting the cells utilise different fatty acids and lipids to keep the cell stable at the higher temperatures.

Perhaps there are even tougher life forms that we are yet to discover, but the thermophilic microorganisms that we have so far identified and investigated in places like Kamchatka all point to the fact that life has its limits. Evolution by natural selection can only adapt so much, and even though it's impossible to imagine what life on Earth will look like in a few hundred million or even a few billion years' time, we know that biology is constrained by thermodynamics, and so we can say with some certainty that there will come a time when the Earth is too hot for any living things to exist. Natural selection will eventually run out of options as the laws of physics outplay it, and all life will come to an end.

Astrobiologist David Grinspoon on Venus as a window on Earth's future

'Left to its own devices, Earth will go the way of Venus. Now, this is nothing to lose sleep over right now because we're talking at least a billion years, probably more like a couple of billion years in the future. We have more immediate concerns, but as we do compare the planetology and look at the exoplanets around other stars and consider the variety of planets in the universe and consider not just the past, but the future of our climate in our solar system, it is something to think about, that the current state of Venus is probably some kind of a window into the distant future of Earth under the warming sun.'

When this will happen no one can be certain, but as the Sun ages and grows hotter, temperatures on Earth will rapidly rise. Today the average surface temperature on the planet is 14.9 degrees Celsius, but with just a 10 per cent rise in the Sun's luminosity, the average temperature will rise to 47 degrees Celsius and climbing. The increased temperatures will raise great storms across the planet. The rains will remove carbon dioxide from the atmosphere and it will be locked away as newly formed sedimentary rock. Trees and plants will struggle as they are robbed of the gas that sustains them, until eventually photosynthesis will cease. The lungs of our planet will fail and the precious oxygen that green plants and algae produce will dwindle. With the primary food source gone, the food chain will collapse and the age of complex life on Earth will draw to a close.

Heat-loving extremophiles may flourish for millions of years more, but eventually nuclear physics will have its way and as average temperatures race above 100 degrees Celsius, the last pockets of life will be extinguished from the Earth.

We can say with confidence this is going to happen because we can plot the future of our Sun far more precisely than the future of the Earth. Our understanding of nuclear physics allows us to predict what happens inside the cores of stars and thus we can see the past, present and future of stars like ours written across the night sky.

The heavens are filled with shining examples of stars that give us a glimpse into the future of our Sun. Arcturus, for example, in the constellation Boötes, is one of the brightest stars in the Northern Hemisphere. It's around the mass of the Sun, perhaps a little bit heavier, and so in the distant past would have had remarkably similar characteristics to our own star. Today, though, Arcturus is 6 to 8 billion years old, potentially 3 billion years older than the Sun, and as it is no longer a main-sequence star, it is now in the red giant phase. Its fuel exhausted, it has swollen up to 25 times its original diameter and is around 170 times as luminous, despite the fact that as its core slowly burns out it is cooling.

To see even further into the future, we need to look towards the brightest star in the northern sky – Sirius. The dog star, as it is commonly known, is twice the mass of the Sun and still fully in the main sequence. But obscured by the glare of Sirius A is a faint companion, Sirius B. This is a star that has already burnt through its fuel, swollen into a red giant and the outer layers have drifted off into space, leaving the fading core of the star about the size of the Earth, known as a white dwarf.

These stars are just two examples amongst many that point us towards the ultimate fate of our Sun, a fate that we believe will play out over the next 5 billion years or so.

Just like Arcturus, as the Sun exhausts its hydrogen fuel, its outer edge will inflate and it will enter a red giant phase. Expanding millions of kilometres out into space, it will engulf Mercury first. Venus's fate will be sealed next as the Sun expands further. Some models predict that Earth may just escape the fiery end of its neighbours – heated to 1,000 degrees Celsius but hanging on beyond the edge of the dying star as its orbit extends out due to the lessening mass of the Sun. Dead but not destroyed, Earth and Mars will orbit as burned-out relics of their former selves. The era of the four rocky inner planets will be over, the billions of lives lived on the surface of one of them nothing but a distant memory, but within our Solar System lies another family of rocky worlds whose moment in the Sun may be to come.

A NEW HOPE

Far beyond the asteroid belt, millions of miles away from the sun-drenched planets of the inner Solar System, the gas giants of Jupiter and Saturn are home to another family of rocky worlds. Jupiter alone has 79 known moons orbiting it, a menagerie of satellites of multiple shapes and sizes. We've been peering at these moons since Galileo Galilei spotted four of them (Io, Europa, Ganymede and Callisto, known as the Galilean moons) over 400 years ago, with his telescope, transforming our understanding of our place in the Solar System.

Today we have explored the Galilean moons not just from afar but close up and found them to be dynamic worlds. Io is fiercely volcanic and Europa, the ice moon, shows tantalising evidence on its surface pointing to a sub-surface ocean sitting below its icy crust. Ganymede and Callisto make up the final two Galilean moons, and just like Europa they are rocky worlds with an abundance of water ice on their surfaces and perhaps their own oceans lurking beneath. These three rocky, frozen worlds are all sitting in the cold outreaches of our Solar System, touched by the distant Sun but barely warmed, lying dormant until perhaps one day the ageing Sun will reach out and turn these bodies into ocean worlds for the very first time.

The next planet out, Saturn, also has its ever-growing family of moons. Amongst its collection of over 60 confirmed satellites are Titan, the only known moon with a dense atmosphere and liquid lakes on its surface (though they are primarily methane, not water), and Enceladus, a frozen ice moon just like Europa with a liquid ocean deep beneath its ice. We will come to Enceladus in detail in Chapter 4, but for now it's intriguing to note that this icy moon may be our best current candidate as a second life-sustaining world in our Solar System. Until we go back and explore further we can't be certain what lies below its surface, but the possibilities that the Cassini probe has so tantalisingly hinted at make it one of the most exciting places for us to visit within the next generation of interplanetary expeditions.

All these ice worlds, sitting dormant in the frozen reaches of the Solar System, offer the promise of a very different future, one in which the rocky worlds of the inner Solar System have been reduced to cinders, and a new generation of worlds waits to awaken. Ice worlds will become water worlds, warmed by the expanding Sun, until our dying star ultimately collapses into a white dwarf.

'The world is my country,
science is my religion.'
Christiaan Huygens

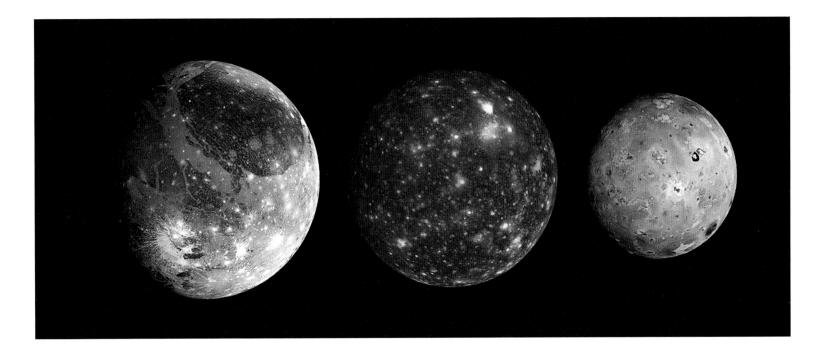

Opposite: Created by images taken by the Galileo spacecraft in the late 1990s, this colour view shows Saturn's icy moon Enceladus – perhaps our closest candidate for sustaining life as we know it.

Above: From left to right, the moons of Jupiter – Ganymede, Callisto and Io – are dynamic worlds; the former two lie dormant, waiting to be awakened by the warmth of the Sun.

Top: Titan, a frozen moon shrouded in its own atmosphere, as seen from Saturn.

EARTH

MARS

THE TWO SISTERS

PROFESSOR BRIAN COX

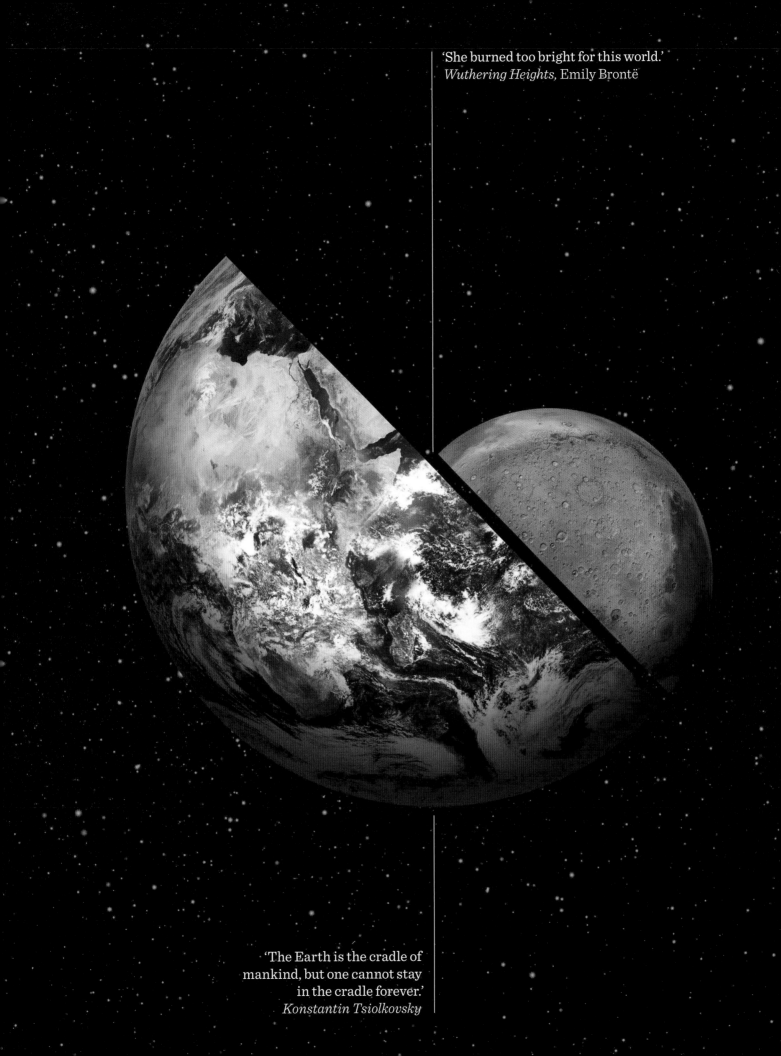

'She burned too bright for this world.'
Wuthering Heights, Emily Brontë

'The Earth is the cradle of
mankind, but one cannot stay
in the cradle forever.'
Konstantin Tsiolkovsky

WAR OF THE WORLDS

Mars is a mirror for our dreams and nightmares. To the naked eye, the planet exhibits a reddish hue, blood red in the imagination; God of War, Star of Judgement. Through a small telescope, it is the most Earth-like of planets, with cinnabar deserts and white polar ice caps. A world we could imagine visiting, perhaps even settling in. Nineteenth-century astronomers convinced themselves they saw plains and mountain ranges and canals delivering meltwater from high latitudes to arid equatorial cities. Some thought the Martians a peaceful civilisation, far in advance of our own. Others saw threat. 'Across the gulf of space, minds that are to our minds as ours are to those of the beasts that perish, intellects vast and cool and unsympathetic, regarded this earth with envious eyes,' wrote H.G. Wells in his classic science-fiction novel *The War of the Worlds*, in 1897.

The nature of Mars remained a mystery until well into the twentieth century because the planet is small and far away and therefore difficult to view with ground-based telescopes. Even the Hubble Space Telescope, high above the distorting effects of Earth's atmosphere, produces images which would not at first sight have prevented Wells from publishing. With a little imagination, the ice caps, high clouds and dark regions circling the deserts could be mistaken for evidence of a water cycle feeding the seasonal advance and retreat of vegetation.

Photographs from the first flyby of Mars by NASA's Mariner 4 spacecraft on 15 July 1965 abruptly laid to rest the romantic notion of Mars as Earth's habitable twin or potential foe. These images revealed an arid surface reminiscent not of our blue planet but of our desiccated Moon. Overnight, we discovered for certain that Earth is the only planet in the Solar System capable of supporting complex life, and contemporary accounts of the impact of the Mariner 4 flyby suggest that this was a powerful realisation. In November 1965, the *Bulletin of the Atomic Scientists* carried an article entitled 'The Message From Mariner 4' – and the message was bleak. 'The shock of Mariner's photographic and radiometric reports is caused not only by their denial of the terrestrial image of Mars, but by the revelation that there is no second chance, at least not in the solar system.' President Lyndon B. Johnson was reported as commenting, 'It may be – it may just be that life as we know it, with its humanity, is more unique than many have thought.' The hesitation in the first few words is revealing. Here is Mars as a symbol of our cosmic isolation. It is as though deep, or perhaps not so deep, in the subconscious, the 1960s' power brokers all the way up to the President suddenly understood that the Earth is far more fragile and precious than a dispassionate analysis of their Cold War brinkmanship might suggest. Or perhaps the perspective delivered by exploration is always shocking. Apollo 8's Earthrise, the photograph that delivered such a positive end to a troubled 1968 by setting the blue Earth against the grey Moon, was three years away, but red Mars provided a foretaste.

Above: The topography of Mars, as captured by NASA's Hubble Space Telescope. The white ice clouds and orange dust storms characterise the planet's hostile weather systems.

Opposite: The Mariner 4 spacecraft began its historic journey to Mars on 28 November 1964. It sent back its first pictures to NASA's Jet Propulsion Laboratory on 15 July 1965.

'The flight of Mariner 4 will long stand as one of the really great advances in man's unending quest to extend the horizons of human knowledge.'
Lyndon B. Johnson

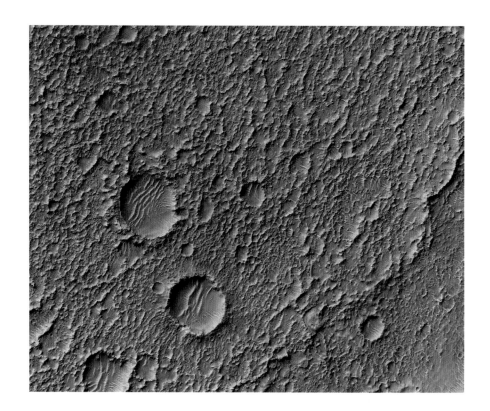

'... if there were intelligent life on Mars
... a photographic system considerably
more sophisticated than Mariner
4 would be required to detect it.'
Carl Sagan

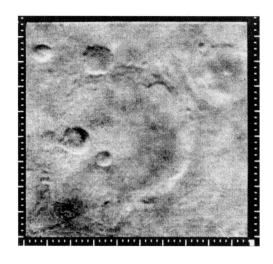

Lyndon B. Johnson – extracts from Remarks Upon
Viewing New Mariner 4 Pictures From Mars,
29 July 1965
'Dr Webb, Dr Pickering, Dr Leighton, Members of
Congress, distinguished guests:

    Unaccustomed as I am to welcoming men from Mars,
I am very happy to see you gentlemen here this morning.
As a member of the generation that Orson Welles scared
out of its wits, I must confess that I am a little bit
relieved that your photographs didn't show more signs
of life out there ...

    The flight of Mariner 4 will long stand as one of the
really great advances in man's unending quest to extend
the horizons of human knowledge. In the history books
of tomorrow, unlike the headlines of today, the project's
name may be lost but the names of the men of vision,
men of imagination and faith who made this enterprise
such a historic success are going to be honored in the
world for many generations to come.

    This advance for mankind is awe-inspiring. It is all
the more so when we realize that such capabilities have
come into being within a short span of a very few years ...

    It may be – it may just be that life as we know it with
its humanity is more unique than many have thought,
and we must remember this.'

Opposite: One site on Mars
seen three ways. First imaged
in 1965 by Mariner 4 (left), then
in 2017 by the HiRISE camera
on the Mars Reconnaissance
Orbiter (MRO) (top right). The
bottom image was hand-
coloured by NASA employees
based on data transmitted
back by Mariner 4.

Above: President Lyndon
B. Johnson sharing in man's
'quest to extend the horizons
of human knowledge' as
Dr William H. Pickering (left),
director of the Jet Propulsion
Laboratory, shows him the
first Mariner 4 photos.

In response, Carl Sagan co-authored a paper suggesting, somewhat playfully, that all was not lost. Mariner 4 took only 22 photographs with a resolution of over a kilometre in a strip crossing the region in which the astronomer Percival Lowell had sketched canals from his observatory in Flagstaff, Arizona, at the turn of the twentieth century. From the warmth of the Arizona Desert, Lowell wrote that Mars was chilly, but no more so than the South of England, which certainly supported a civilisation of sorts. Using several thousand photographs of a similar resolution taken by meteorological satellites in Earth's orbit, Sagan and his co-authors found only a single feature that unambiguously indicated the presence of a civilisation – Interstate Highway 40 in Tennessee. They concluded that Mariner 4 would not have detected human civilisation had it flown by Earth. 'We do not expect intelligent life on Mars, but if there were intelligent life on Mars, comparable to that on Earth, a photographic system considerably more sophisticated than Mariner 4 would be required to detect it.'

The non-existence of an extant Martian civilisation was confirmed by the Mariner 9 mission in November 1971 – the first spacecraft to orbit another planet. Mariner 9 achieved a photographic resolution of 100m per pixel, and no sign of intelligent life, past or present, was detected. The twin Viking landers in 1976 failed to detect even microbial life, although the combined results of the suite of microbiology experiments carried by the spacecraft are not considered to be unequivocal because Martian soil chemistry is, to coin a phrase from the official NASA report, enigmatic, and could conceivably have masked any biological activity.

In hindsight, the fact that Mars is not teeming with life today is not so surprising. Mars orbits 50 million miles further from the Sun than Earth and receives less than half the solar energy. It is a small world with a tenuous atmosphere that provides little insulation or greenhouse warming. NASA's Curiosity rover in the Gale crater has measured midday temperatures above 20 degrees Celsius, but in the early hours of the morning it has experienced minus 120. As Alfred Russel Wallace wrote in 1907, any attempt to transport water across the Martian surface today would be 'the work of mad men rather than of intelligent beings'. There are no canals, no cities and no envious eyes. The planet is a frozen hyper-arid desert too far from the Sun to support complex life.

Yet it hasn't always been this way. Observations from our fleet of orbiting spacecraft and landers have revealed a complex and varied past. Once upon a time the red planet was glistening blue. Streams ran down hillsides and rivers wound their way through valleys, carved by a water cycle from land to sky and down again from mountains and highlands to the sea. This presents a great challenge for planetary scientists. Put simply, nobody would have been surprised if Mars had always been an inert rock because it is a small planet far from its star. But the geological evidence is unequivocal; the surface tells a different story.

Mars, then, remains an enigma. As a wandering red star, it stirred the imagination of the ancients. As a telescopic image, too small and shifting for visual or intellectual clarity, it became our twin. When spacecraft flew by, it shocked us into considering our cosmic isolation. The red planet was relegated in our collective consciousness to the status of just one more rock glistening in the night. Then we landed, and discovered a world that was once habitable, and could be again.

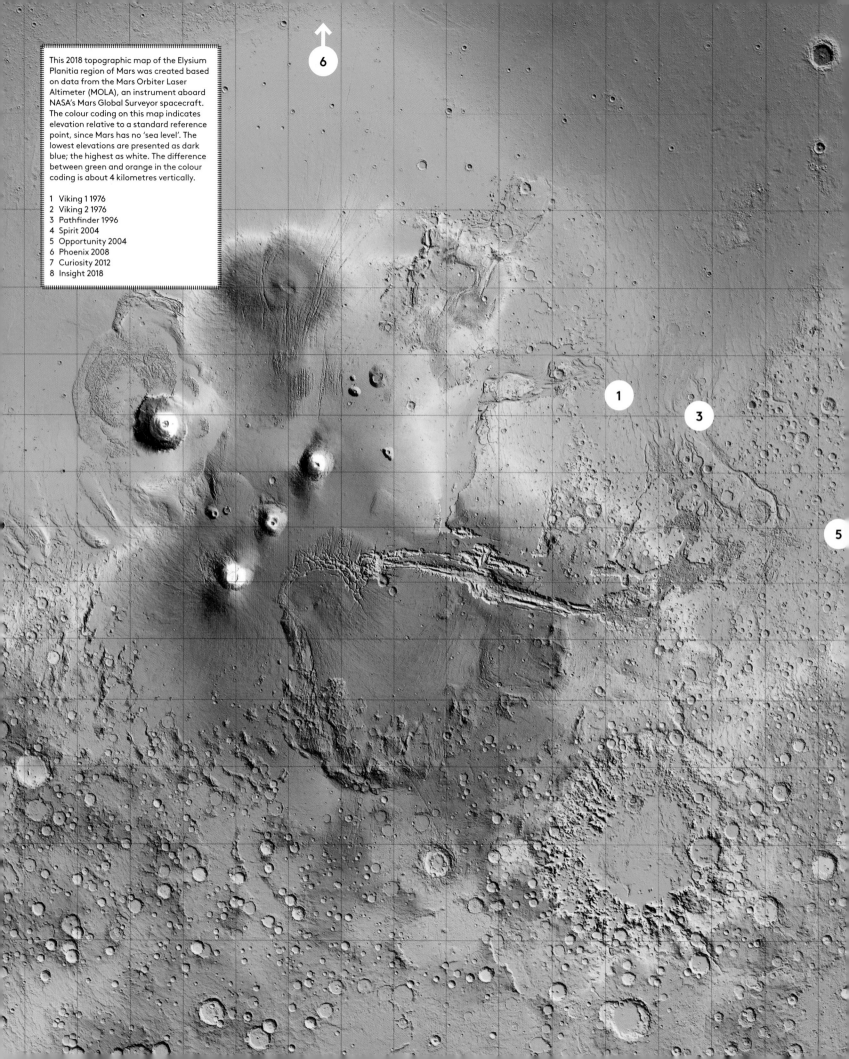

This 2018 topographic map of the Elysium Planitia region of Mars was created based on data from the Mars Orbiter Laser Altimeter (MOLA), an instrument aboard NASA's Mars Global Surveyor spacecraft. The colour coding on this map indicates elevation relative to a standard reference point, since Mars has no 'sea level'. The lowest elevations are presented as dark blue; the highest as white. The difference between green and orange in the colour coding is about 4 kilometres vertically.

1 Viking 1 1976
2 Viking 2 1976
3 Pathfinder 1996
4 Spirit 2004
5 Opportunity 2004
6 Phoenix 2008
7 Curiosity 2012
8 Insight 2018

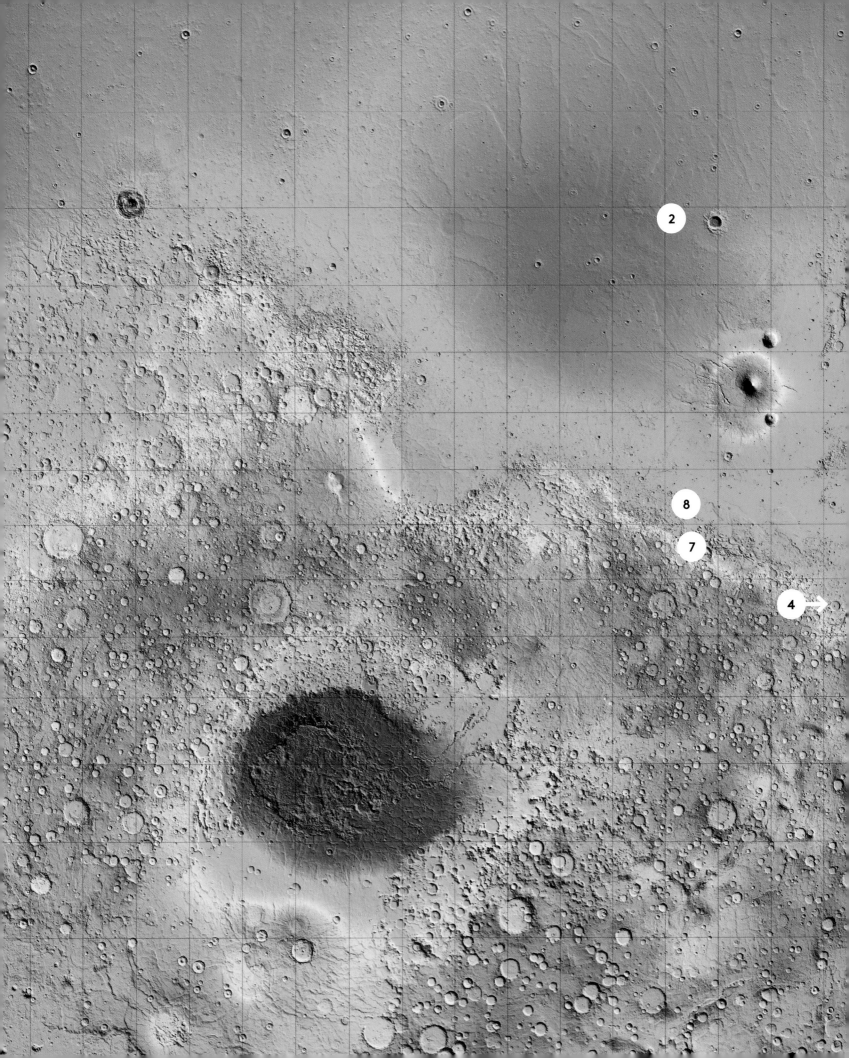

READING THE MAPS OF MARS

'We found three-and-a-half-billion-year-old lake sediments that contained a diversity of organic materials. It tells us there's actually organic matter present. What we found in those rocks, is what we expected of natural organic matter. It's what you would expect to find on Earth.'
Jennifer Eigenbrode, astrobiologist

A map of Mars can be read like a history book. Unlike Earth, where constant weathering, tectonic activity and volcanism have erased the deep geological past, Mars has been relatively quiescent for most of its life. The scars of collisions from the first turbulent billion years after the formation of the Solar System can still be seen from orbit; ancient cataclysms documented below a thin film of dust.

NASA's Mars Global Surveyor spacecraft spent four and a half years mapping Mars in the late 1990s and provided detailed maps such as the one shown on the previous page, with colours corresponding to differences in altitude. Just as on Earth, there is significant variation, but the geological features on our smaller sister world are much bigger and bolder.

The highest elevations on Mars are found on the Tharsis Rise, a great volcanic plateau and home to the largest volcano in the Solar System, Olympus Mons. At over twice the height of Everest, Olympus Mons towers 25 kilometres above the lowlands of Amazonis Planitia to the west, and its base would fit inside France, just about. Cutting a deep scar across Tharsis to the south-east of Olympus Mons is Valles Marineris, named after the Mariner 9 spacecraft that discovered it, a canyon that dwarfs anything on Earth; the Grand Canyon would fit into one of its side channels.

The lowest points on Mars are found in the Hellas impact basin, the largest clearly visible impact crater in the Solar System. From the highest points on the crater rim to the floor, Hellas is over 9 kilometres deep; it could contain Mount Everest. The atmospheric pressure at the floor is twice that at the rim; high enough for liquid water to exist on the surface in a narrow range of temperatures.

These are extreme altitude differences for a small world; over 30 kilometres from the summit of Olympus Mons to the floor of Hellas. On much-larger Earth, for comparison, there is only 20 kilometres difference between the summit of Everest and the Challenger Deep in the depths of the Mariana Trench.

The most striking and ancient elevation difference on Mars is that between the Northern and Southern Hemispheres of the planet, known as the global dichotomy; Mars is an asymmetric world. The Northern Hemisphere is on average 5.5 kilometres lower in altitude than the Southern. There is no consensus as to how the dichotomy formed, other than that it was early in the planet's history and before the large impacts which created the Utopia and Chryse Basins around 4 billion years ago. At some later time, the Northern lowlands were resurfaced by volcanic activity in a similar fashion to the smooth lunar seas, which accounts for their lack of cratering relative to the much more ancient terrain to the south.

The oldest terrain on Mars is found in the Noachis Terra region of the Southern Highlands. It is characterised by heavy cratering reminiscent of the far side of the Moon. Even small craters in the Noachian Highlands are heavily eroded, which suggests the regular, if not persistent, presence of liquid water. There are dry river valleys and deltas and evidence of water pooling in the craters and overflowing their walls, forming interconnected networks of lakes. This is how we know Mars was once a warmer and wetter world, at least occasionally; the evidence is written across the Land of Noah.

In contrast, the younger terrain of Hesperia Planum displays much less evidence of regular erosion by water, but bears the scars of occasional catastrophic floods that cut deep valleys over very short periods of time and may have formed temporary large lakes or seas.

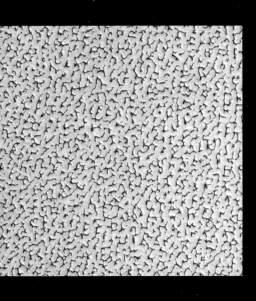

Top left: Winter frosts at the North Pole of Mars are disappearing, revealing the surface features of the ice cap.

Bottom: Geological faults have disrupted layered deposits, creating a striking landscape in the northern Meridiani Planum region.

Top right: Alluvial fans are gently sloping wedges of sediment deposited by flowing water. Some of the best-preserved examples on Mars are in Saheki crater. On Earth, they are found in deserts, for example, in Death Valley, California.

Opposite: The Grand Canyon, in Arizona, began to form about 1,200 million years ago in the late Proterozoic period. Mars was already long frozen by this point.

The Amazonis Planitia region shows little sign of flowing water, fewer impact craters and less evidence of active volcanism, suggesting it was formed more recently when Mars was significantly less geologically active.

The persistence of surface features over many billions of years in the Noachian, Hesperian and Amazonian regions has led to the historical epochs of Mars being named after the distinctive terrains that still bear the characteristic marks of the climate and geological activity that formed and sculpted them.

The Noachian Period was the earliest and wettest, and coincided with the origin of life on Earth around 4 billion years ago, when conditions on both worlds appear to have been very similar. The Martian atmosphere may have been denser than Earth's, and dominated by carbon dioxide, but significant questions remain about how such an atmosphere could have warmed Mars sufficiently to deliver the warm, wet climate and how that atmosphere was lost. The MAVEN spacecraft currently in orbit around Mars aims to answer this question, as we'll discuss later in this chapter. The Noachian Period ended as Mars became increasingly cold and arid around 3.5 billion years ago, just as life was gaining a foothold on Earth.

The Hesperian Period, the time of catastrophic floods, ran from the end of the Noachian to around 3 billion years ago, when Mars entered its current frozen, arid phase, punctuated by occasional volcanic activity and the large-scale movement of ice, but with very little evidence of flowing water. The long 3-billion-year freeze from the end of the Hesperian to the present day is known as the Amazonian.

This is a summary of what we know about Mars; the whys pose a significant challenge to planetary scientists. Given a warm, wet and seemingly stable world early in its history, what triggered the loss of atmosphere and descent into modern-day aridity? What happened to the water on Mars? Was it lost to space or does it persist today as surface ice or in subsurface rocks or reservoirs? If so, how much water is still accessible? Could we exploit the ancient reservoirs of Mars to support a human colony? And perhaps most significantly of all, did life arise on the planet during the Noachian Period, coincident with the origin of life on Earth, and could that life still be present on Mars today?

The current fleet of spacecraft in orbit around Mars and roving across its surface has been designed to answer these questions.

Right: The history of Mars from formation to the present, including major geological events, is shown in comparison to Earth's timeline. Time is measured in Ga (*giga annum*): billions of years before the present. The Phanerozoic and Amazonian eons extend to and include the present day on the two planets.

THE RELATIVE TIMELINES OF LIFE ON EARTH AND ACTIVITY ON MARS

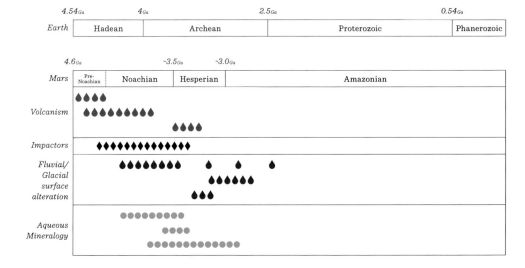

THE MARTIAN FLEET

'It may be — it may just be that life as we know it with its humanity is more unique than many have thought, and we must remember this.'
President Lyndon B. Johnson

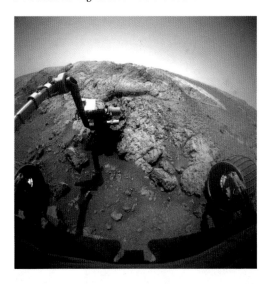

Mars today is a planet buzzing with activity. Communications to Earth and the Martian Internet are managed by the Mars Reconnaissance Orbiter (MRO), an orbiting bridge between worlds. MRO carries the HiRISE instrument, a camera with resolution high enough to see basketball-sized features on the Martian surface. The Mars Color Imager (MARCI) camera monitors Martian weather, and the Compact Reconnaissance Imaging Spectrometer for Mars (CRISM) identifies mineral deposits, particularly those formed in the presence of surface water.

Orbiting with the MRO is the Mars Atmosphere and Volatile Evolution Mission (MAVEN). This camera-less spacecraft operates between 150 kilometres and 6,000 kilometres above the Martian surface, measuring the composition of the atmosphere at different altitudes and observing how the tenuous gases are stripped from the planet by the solar wind.

Mars Odyssey is the veteran of the orbiting fleet, having arrived in 2001 and still being operational in a polar orbit, searching primarily for water ice on the surface. Mars Express is a European Space Agency mission that is delivering high-resolution photographs, mineralogy data, radar investigation of the near sub-surface and atmospheric measurements, including the search for methane, a gas that on Earth is associated with biological activity. India's Mangalyaan space probe is primarily a technology demonstrator, but it carries a secondary scientific package capable of investigating atmospheric composition.

The newest arrival at Mars is the joint European Space Agency/Russian ExoMars Trace Gas Orbiter, which will observe seasonal changes in the Martian atmosphere and search for subsurface water deposits. The spacecraft will form the communications bridge for ESA's ExoMars rover, due to land in 2021.

The two most recent explorers of Mars are the Opportunity and Curiosity rovers. The Opportunity rover landed on the Meridiani Planum close to the Martian equator on 25 January 2004, with a planned lifetime of 90 Earth days. In a spectacular testament to the Jet Propulsion Laboratory's engineering excellence, Opportunity remained operational until a planet-wide dust storm covered its solar panels in June 2018, after over 14 years and a journey of 45 kilometres on the surface of Mars, exploring the Endurance, Victoria and Endeavour craters. On 13 February 2019, Opportunity was finally declared 'dead'.

Opportunity's younger and far larger companion on Mars is Curiosity, the most massive and most capable spacecraft ever to touch down on a planet beyond Earth. The landing itself was a tour de force of engineering ingenuity and audacity. The enormity of the mission is best described through the words of Allen Chen, Operations Lead at NASA for the Curiosity Mission, at 10.31 PDT on 5 August 2012, watching the landing at the Jet Propulsion Laboratory alongside the team (see overleaf).

The engineers described it as seven minutes of terror. Seven minutes to manoeuvre a spacecraft that had taken eight years to design and build, with a programme cost of more than $2.5 billion, from 13,000 miles per hour at the top of the Martian atmosphere to a soft landing on the surface of the planet. The tension reflected the high stakes; the Mars Science Laboratory mission, to give the lander and orbiter their full title, was a high-risk flagship science mission in the same category as Voyager, the Hubble Space Telescope, Viking and Cassini, and was beset by cost and schedule overruns and controversy. This is in many ways unsurprising – new

Opposite: An artist's illustration of NASA's Mars Reconnaissance Orbiter (MRO) passing above Nilosyrtis Mensae, a portion of the planet.

Above: An Opportunity eye view of the surface of Mars, taken from the rover's front hazard-avoidance camera (Hazcam).

Seven Minutes of Terror – in the words of Curiosity
mission's Allen Chen
Things are looking good. Coming up on entry
Vehicle reports entry interface
We're beginning to feel the atmosphere as we go in here
Alright, it is reporting that we are seeing Gs on the
    order of 11 or 12 Earth Gs
Bank reversal 2 is starting
We are now getting telemetry from Odyssey
We should have parachute deploy around Mach 1.7
Parachute has deployed
We are decelerating
Heat shield has separated, we are locked on the ground
We're down to 90 metres per second at an altitude
    of 6.5 kilometres and descending
Standing by for backshell separation
We are in powered flight
We are at altitude 1 kilometre and descending
Standing by for sky crane
Sky crane is starting
Signal from Odyssey remains strong
Touchdown confirmed! We're safe on Mars!
(cheering, applause)
We got thumbnails. It's a wheel! It's a wheel!

technologies and ambitious scientific objectives are difficult to implement and achieve, and in part because of this deliver great rewards.

In hindsight, nobody would question the value of any of these missions, which have delivered some of the greatest insights and most inspiring images in the history of exploration. Yet this never prevents bean counters and rival scientists with agendas and, more charitably, budgetary challenges of their own, whingeing and even considering cancellation. This is naïve; funding for science is almost always grudging from a political class whose view of the acquisition of knowledge is utilitarian. The far deeper value of exploration as a critical part of the internal voyage of our species, bringing us into direct confrontation with the mystery of our existence, is lost on them, at least until occasional Johnsonian (President, not Alexander Boris de Pfeffel) moments: 'It may be – it may just be that life as we know it with its humanity is more unique than many have thought, and we must remember this.' Which means that a cancellation of one project does not mean an uplift in funds for another. More likely the budget will be lost to science.

At the time of writing the same issues plague another NASA flagship mission, the James Webb Space Telescope. As Robert D. Baum wrote in an OpEd article for *Space News* in December 2008, at the height of the Mars Science Laboratory controversy, 'When implementing flagship missions, cost and schedule overruns are not uncommon, but history shows that the mission return often eclipses the expenditure. The Hubble Space Telescope experienced a development cost overrun several times its approved project budget and was launched much later than originally planned. In hindsight, would any rational space scientist not concede that the return from Hubble was worth the investment?'

Given the febrile atmosphere, however, no amount of philosophy or reason would mitigate the costs of failure of the Mars Science Laboratory mission; a landing accident was both possible and unthinkable. Delivering almost a tonne of fragile rover onto the surface of Mars with pinpoint accuracy is not trivial, which is perhaps why the chosen engineering solution seemed, not to put too fine a point on it, daring.

Above: Curiosity rover and its parachute descending to the surface, photographed from space by the HiRISE camera on the MRO.

Right: Celebration at the Jet Propulsion Lab after watching Curiosity rover land safely on Mars.

OFF-PLANET DRIVING RECORDS

■ *Lunar Rover* ■ *Mars Rover*

Opportunity
2004–2019

45.16 km

Lunokhod 2
USSR
1973

37 km

Apollo 17
Lunar Rover
1972

35.74 km

Apollo 15
Lunar Rover
1971

27.8 km

Apollo 16
Lunar Rover
1972

26.7 km

Curiosity
2012–present

22.22 km

Lunokhod 1
USSR
1973

10.5 km

Spirit
2004–2010

7.7 km

Sojourner
1997

0.1 km

Mars has a thin atmosphere, so slowing down a spacecraft when it's travelling around ten times faster than a bullet is hard. And yet the atmosphere is thick enough to apply sufficient frictional heating to destroy a spacecraft without adequate protection, and turbulent enough to cause significant uncertainty in the landing site of an unsteered vehicle. This renders the more obvious engineering solutions redundant; you can't simply deploy a parachute and float to the surface.

The 1970s' Viking landers used a combination of heatshields during aerobraking, parachutes and retrorockets with success, but the Curiosity engineering team dismissed this tried-and-tested solution; it did not allow for high enough precision, or for the gentle landing required for the pinpoint delivery of a large rover with delicate wheels; the Viking landers had legs built like tanks. The Viking landers also carried their heavy descent rocket systems with them to the ground, which didn't matter because they stayed put. Curiosity would have had to lug all this unnecessary mass around Mars for years.

The Opportunity rover, her sister ship Spirit and the earlier Mars Pathfinder employed a combination of aerobraking, parachutes, rockets and airbags to cushion the falling spacecraft, but these previous rovers were lightweight in comparison, at just over a fifth of the mass of the gargantuan Curiosity. A rover of this size and complexity bouncing over the surface in an airbag cocoon was not considered feasible, primarily because of the sheer mass of the airbag system needed to cushion the impact; the lightweight Opportunity rover's landing system was almost twice the mass of the rover itself.

The chosen Entry, Descent and Landing procedure (EDL) for Curiosity was described by mission lead Adam Steltzner as the 'result of reasoned, engineering thought'; a beautiful example of the truism that reasoned engineering thought and common sense, at least of the sort possessed by the average golf-club bore, do not always match up. The Curiosity EDL appears to the untrained eye to be bonkers, or maybe overcomplicated, but it wasn't, and it worked.

Above: This chart compares the distances driven by various wheeled vehicles on the surface of Earth's Moon and Mars. Of the vehicles shown, only the NASA Mars rover Curiosity is still active. The figure shows their distances driven as of 26 February 2019.

Right: On 15 August 2014, Opportunity beamed back this panoramic image of its tyre tracks while exploring the western rim of the Endeavour crater. If you look closely, you can see its tracks extend down towards Murray Ridge from earlier in 2014.

'When Curiosity had landed safely
I felt a huge sense of relief; melting
in your chair knowing that we
had successfully landed on Mars.'
Ashwin Vasavada, planetary scientist

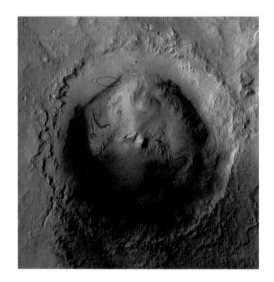

Above: The target landing area
for Curiosity rover is the ellipse
marked on this image of Gale
crater. The ellipse is about
20 km by 7 km.

Tucked up inside a protective shell, Curiosity encountered the outer layers of the Martian atmosphere 125 kilometres above the surface and travelling at around 20,000 km/h at 10.23 pm Pacific Daylight Time on 5 August 2012. The spacecraft, at this stage cocooned inside a protective enclosure, flew entirely under the control of its onboard computers with no input from mission control on Earth. This was the first time an 'autonomous guided entry system' was employed for an interplanetary mission, using a combination of thrusters and the ejection of 'balance masses' to shift the centre of mass and trajectory of the spaceship and guide it with a design accuracy of around 10 kilometres towards the chosen landing site. This 'guided', rather than 'ballistic', trajectory allowed for a much wider choice of landing sites for Curiosity than for the previous landers, which required around 100 kilometres of flat, safe land surrounding the nominal touchdown point. But it also required the rover to fly itself. The round-trip travel time for a radio signal to Mars was over 13 minutes when Curiosity landed, making control from the Jet Propulsion Laboratory in California impossible.

During the first stage of descent a heat shield constructed from a unique material known as PICA (phenolic impregnated carbon ablator) protected Curiosity from peak temperatures of around 2,000 degrees Celsius generated by friction in the thickening Martian atmosphere. Four minutes after beginning the entry phase, slowed to just under 3,000 km/h, Curiosity's parachute deployed at an altitude of 11 kilometres above the surface. The supersonic parachute is a vast and complex structure; 80 suspension lines over 50 metres long attached to a 16-metre canopy.

Having slowed to a velocity of 700 km/h with the help of the parachute, the heat shield was jettisoned at an altitude of 8 kilometres, allowing the onboard radar to get a view of the ground and deliver high-precision altitude and velocity measurements. With the parachute fully unfolded in relatively dense, low-level Martian atmosphere, Curiosity drifted in the Martian sky for 80 seconds, descending ever more slowly.

At an altitude of 1.8 kilometres, travelling at a sedate 280 km/h, the NASA timeline calls for a 'deep breath' as Curiosity separated from the parachute system and dropped in freefall towards Mars. 'For engineers,' says the NASA website, 'it's like jumping out of a plane for the first time.' The freefall drop was designed to allow Curiosity to separate far enough from the parachute system so that it wouldn't accelerate back into it when, 300 metres further down, its retro-rockets fired and the powered descent phase of the landing began. For the Viking landers, this was the final phase, but for Curiosity, after the rockets had delivered the rover to an altitude of just 20 metres with a descent velocity of less than 1 m/s – almost a hover – the audacious final phase of the landing protocol kicked into action: the Sky crane.

The rover slowly departed the rocket cradle, attached by three nylon wires and an umbilical cord of electrical connections 7.5 metres in length. Four of the rockets, angled away from the vertical so as not to damage the rover, continued to fire, allowing the whole delicate system to approach the surface at walking pace. At 5 metres from the surface, Curiosity unfurled its wheels, stowed for the eight-month journey from Earth. At 10.32 pm, the rover confirmed its wheels were in contact with the Martian surface, and issued one last command through the electrical umbilical up to the Sky crane: 'I'm down. Cut the cables and fly away.' After a half-a-billion-kilometre journey, Curiosity was delivered safely to Mars just 2.4 kilometres from its nominal landing point. A triumph of engineering ingenuity and brilliance.

Landing Curiosity took the largest parachute ever built, which was extensively tested inside the world's largest wind tunnel, at NASA Ames Research Center in California.

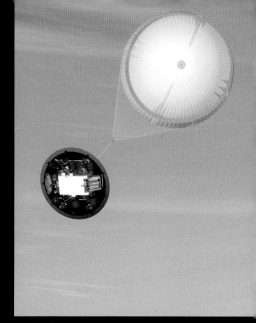

Left: About the size of a small SUV, NASA's Curiosity rover has six-wheel drive and the ability to turn a full 360 degrees, as well as the agility to climb steep hills.

Above: The final 'sky crane' stage of Curiosity's descent lowered the rover to the planet's surface.

THE MARS CURIOSITY ROVER

*Length: 3 m; Width: 2.7 m; Height: 2.2 m; Mass: 899 kg.
Top speed of 4 cm per second on hard, flat ground.
Capable of driving over 19 km.*

**Traveller Eyes Navigation
Cameras (Navcams)**
Aids in autonomous navigation

**Laser Eye Chemistry &
Camera (ChemCam)**
*Analyses chemical composition
using telescope, remote micro-
imager, laser, spectrometer*

Wheels
*Made of aluminium, with
curved titanium springs for
springy support. At 50 cm
in diameter, one full turn
of a wheel with no slippage
covers about 157 cm*

Tail (Power Source)
*Uses 4.8 kg of plutonium
dioxide to provide a steady
supply of heat. Slightly
over 100 watts of electrical
power produced*

Legs
*Made from titanium tubing.
Capable of rolling over rocks
the size of one of its wheels*

**Human-like Eyes Mast
Camera (Mastcam)**
*Colour Stereo Imaging similar
to that of consumer digital
cameras; 2 megapixels. 10
frames per second HD Video*

**Water-finding Sense
Dynamic Albedo of
Neutrons (DAN)**
*Measures subsurface hydrogen
up to 1 m below the surface.
Can detect water content as
low as one-tenth of 1%*

Big Mouth UHF Antenna
*Transmits data to Earth
through Mars Orbiters
Radio Frequency: ultra-high
frequency (UHF) band
(about 400 megahertz)*

Sundial
*Additional colour
calibration target*

1 Neck and Head (Mast)
- About 2 metres from the ground
- Cameras give a view from a human perspective and allow remote-sensing capabilities

2 Weather Detector Rover Environmental Monitoring Station
- Two bolt-like booms on the rover's mast measure wind, ground temperature and humidity
- UV sensor on the rover deck, about 1.5 metres above ground
- Designed to survive temperatures from -130°C to +70°C
- All sensors record at least 5 minutes of data at 1 Hz each hour, every Martian day

3 Body
- Covers and protects the computer, electronics and instruments
- The chassis forms the bottom and sides, the top holds the equipment deck

Landing Curiosity: *In order to safely land Curiosity on the surface of Mars, NASA engineers had to get creative...*

Entry Interface
Altitude: 125 km
Velocity: 5,900 m/s
Time: entry + 0 sec

Parachute Deploy
Altitude: 11 km
Velocity: 405 m/s
Time: entry + 254 sec

Heatshield Separation
Altitude: 8 km
Velocity: 125 m/s
Time: entry + 278 sec

Backshell Separation
Altitude: 1.6 km
Velocity: 80 m/s
Time: entry + 364 sec

Sky crane
Altitude: 20 m
Descent velocity -1 m/s

4 Action Eye Mars Descent Imager (MARDI)
- Mounted pointing towards the ground, this takes pictures during descent through the atmosphere
- 8 Gb flash memory allows over 4,000 raw frames
- HD Video: Four colour frames per second (close to 1,600 X 1,200 pixels per frame)

5 Arm
- 2.1 metres in length
- Five rotary actuators: the shoulder azimuth joint, shoulder elevation joint, elbow joint, wrist joint and turret joint enable the hand to work as a human geologist would – grinding away layers, taking microscopic images and analysing the elemental composition of the rocks and soil

6 Hand
- The Collection and Handling for In-situ Rock Analysis (CHIMRA) is a scooping paw that analyses chemical elements in Martian rock and regolith
- The Powder Acquisition Drill System (PADS) is a rotary percussive drill that collects and processes samples for analysis. Diameter of drilled hole: 1.6 cm

CURIOSITY
ON MARS

Curiosity was sent to Mars to explore Gale crater, a 150-kilometre-wide impact crater formed during the late Noachian or early Hesperian Period when liquid water would have been present at least occasionally on the surface. In the great tradition of astronomy, the crater is named after the Australian amateur astronomer, planetary observer, comet hunter and occasional banker Walter Frederick Gale, who discovered a host of comets as well as a number of geological features on Mars at the turn of the twentieth century using self-built telescopes in his back yard.

The primary reason for choosing Gale crater was the unusual central structure, Mount Sharp (or Aeolis Mons), which rises over 5 kilometres above the crater floor. There is still debate about precisely how Mount Sharp formed, but the layers visible from orbit along its flanks suggest that it is the weathered remains of sedimentary rocks which once filled the crater and were laid down over time after

Below: The first sample of powdered rock extracted by Curiosity's drill. The image was taken after the sample was transferred from the drill to the rover's scoop.

Bottom: Curiosity rover looks uphill to Mount Sharp. Spanning the centre of the image is an area with clay-bearing rocks that scientists are eager to explore.

'... The rock record preserved in those layers holds stories that are billions of years old – stories about whether, when, and for how long Mars might have been habitable.'
Joy Crisp, planetary geologist

Above: Curiosity takes a selfie on the lower slopes of Mount Sharp. This is actually a mosaic of images and does not include the rover's arm.

1 Sample Analysis at Mars (SAM)
- Instrument suite
- Identify a wide range of organic (carbon-containing) compounds
- Heat most rock samples to about 1,000°C to extract gases for analysis

2 Chemistry & Mineralogy X-Ray Diffraction/ X-Ray Fluorescence Instrument (CheMin)
- Analyse mineralogy/chemical composition
- An X-ray diffraction and fluorescence instrument

3 Radiation Assessment Detector (RAD)
- Characterise space and Mars surface radiation
- Pointed towards the sky
- Similar size to a small toaster

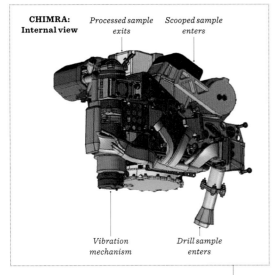

CHIMRA:
Internal view

*Processed sample
exits*

*Scooped sample
enters*

*Vibration
mechanism*

*Drill sample
enters*

SA/SPaH:
Curiosity's Hand

APXS
*Alpha Particle X-Ray
Spectrometer*

CHIMRA
*Collection and
Handling for
In-Situ Martian
Rock Analysis: scoops
regolith, sieves
and portions*

MAHLI
*Mars Hand
Lens Imager*

DRT
*Dust Removal
Tool*

DRILL
*Acquires powder
from rocks*

DRILL:
Top view

DRILL BIT:
Section view

Exit to CHIMRA

Chamber 2

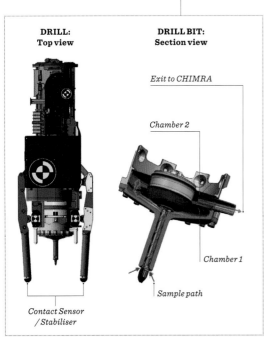

Chamber 1

Sample path

*Contact Sensor
/ Stabiliser*

the impact. Subsequent erosion by the relentless Martian winds removed much of the surrounding rock to reveal the ancient crater floor once more, while leaving the towering central structure intact. Deep, exposed sedimentary layers are extremely enticing to geologists because a cross-section through rock is a cross-section through time. As surface and atmospheric conditions change, different sorts of rock are laid down and chemically modified. As Curiosity ascends the slopes of Mount Sharp, it climbs forwards through Martian time, and scientists are close behind.

Joy Crisp, the Deputy Project Scientist from NASA's Jet Propulsion Laboratory, explained the choice of landing site in the weeks before launch; Mount Sharp 'may be one of the thickest exposed sections of layered sedimentary rocks in the solar system … The rock record preserved in those layers holds stories that are billions of years old – stories about whether, when, and for how long Mars might have been habitable.' Just as the layers of exposed rock in the Grand Canyon on Earth reveal the story of our planet, the exposed sedimentary layers of Mount Sharp are a story book that the winds of Mars have opened, ready for Curiosity to read.

At the time of writing, Curiosity has travelled almost 19 kilometres from its landing site on the flat plains to the lower slopes of Mount Sharp, stopping along the way at each interesting site to characterise the geological environment. Curiosity's suite of scientific instruments is the most sophisticated ever installed on a spacecraft; the rover is a mobile geological laboratory capable of analysis of the Martian surface and the layers just beneath.

NASA loves acronyms. Curiosity acquires samples using the SSS – the Surface Sampling and Science system. It is made up of three parts: the SA/SPaH (Sample Acquisition, Processing and Handling subsystem), SAM (Sample Analysis at Mars instrument) and CheMin (Chemistry and Mineralogy instrument). The main components of SA/SPaH, which is mounted at the end of Curiosity's robot arm, are an integrated scoop and sample-processing system (CHIMRA) and a drill for acquiring surface and subsurface samples. (The scoop and sample processing system surely has one of the most inventive of all NASA acronyms: CHIMRA, pronounced Chimera after the multi-headed creature of Greek mythology, which stands for the Collection and Handling for In-Situ Martian Rock Analysis tool. Someone deserves an award for that.) These are transferred into SAM, which includes a gas chromatograph, a mass spectrometer and a tuneable laser spectrometer, and CheMin, the first X-ray diffraction experiment ever flown in space and the most sophisticated modern technique for characterising mineral samples on Earth.

The science return from Curiosity is only just beginning, as older data are analysed and published and new data from its climb up the slopes of Mount Sharp and through Martian geological time continues. But current results are consistent with and greatly enhance the detail of our picture of a planet that was warmer and wetter when Gale formed around 3.5 billion years ago.

In early October 2012, Curiosity rolled to a halt at a place the geologists christened 'Rocknest' and scooped up a handful of Mars. The sample of dust, dirt and finely grained soil was fed into the SAM and heated to 835 degrees Celsius. The baking process revealed the presence of significant amounts of carbon dioxide, oxygen and sulphur compounds. SAM's analysis also suggested the presence of carbonates, which form in the presence of water.

Left: This diagram illustrates the complexity of Curiosity's hand, the SA/SPaH and its component parts, the CHIMRA (top) and the drill (bottom).

Top: Mount Sharp rises about 5.5 km above the floor of Gale crater. Here the height of Mount Sharp (in elevation above the crater floor) is shown in comparison to the size of large mountains on Earth, whose heights are indicated in elevation above Earth's sea level.

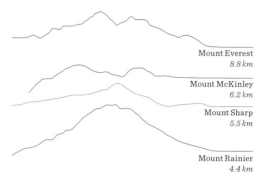

Mount Everest
8.8 km

Mount McKinley
6.2 km

Mount Sharp
5.5 km

Mount Rainier
4.4 km

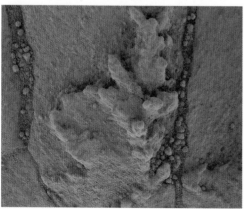

Perhaps most surprising of all, the soil from Rocknest didn't just provide circumstantial evidence for ancient water; around 3 per cent of the sample's mass was water, suspended as small, ephemeral droplets of salty liquid.

Curiosity is sitting on the floor of a lost lake, a body of water that filled Gale at the Noachian/Hesperian boundary. The patterns of sedimentary deposits suggest the lake rose and fell in multiple cycles over tens of millions of years. A transient landscape of ancient streams, deltas and ponds is bounded by the border of the crater. When alighting on ancient river beds running down from the crater's rim, Curiosity has found sulphur, nitrogen, hydrogen, oxygen, phosphorus and carbon – all the necessary building blocks for life. Summarising the results in August 2017, Michael Meyer, NASA's lead scientist for the Mars Exploration Program, said, 'A fundamental question for this mission is whether Mars could have supported a habitable environment. From what we know now, the answer is yes.'

In June 2018, two new discoveries strongly reinforced the idea that Mars was once a habitable world, and maybe still is. Curiosity found complex organic molecules a few centimetres below the surface in rocks known as mudstone that form from silt deposited on lake beds. This means that all the ingredients for life were present at the time when Gale crater was filled with water. Detected molecules include benzene, toluene, propane and butane. Even more tantalisingly, Curiosity also observed a strong seasonal variation in methane levels in the Martian atmosphere today, reaffirming and enhancing previous observations of methane spikes from orbit. The important additional information from Curiosity was that the methane peaked repeatedly in the warm summer months and declined in the cooler winter, which is what would be expected if the methane had a biological origin. This is not irrefutable evidence of life on Mars today by any means – it is thought that geological processes may be able to account for such a seasonal variation, although if this were Earth, a biological explanation would be favoured. The appropriately cautious Michael Meyer had this to say in June 2018: 'Are there signs of life on Mars? We don't know, but these results tell us we are on the right track.' Science is a humble and cautious pursuit, and from the time of Viking we've known that Martian soil chemistry is notoriously complex and can easily catch out the unwary scientist, so we should leave it there.

Curiosity continues its journey through Martian geological history emboldened; it is certainly driving across an ancient lake bed where, for a reasonably extended period over 3 billion years ago, conditions were favourable and all the ingredients for life were present. While we shouldn't speculate or read too much into the results to date, we can let our imaginations wander for a moment to picture what Gale crater might have been like during the late Noachian, when life was gaining a foothold on her sister world.

We stand on the shores of a lake partially filling Gale, fed by runoff from snow melt on the far northern rim. Where Mount Sharp now stands, a small island breaks the surface blue; not the mountain of today but a small central peak left over from the impact. A place of grand beauty where each Martian evening the pale sun sets in the west beyond the island, rays glinting dimly off the still waters interrupted by the shadow of the peak out to the crater's edge. A sundial, marking the passing of time, on a lake populated by microbial Martians.

Middle: This dark mound, called Ireson Hill, rises about 5 m above redder layered outcrop material of the Murray formation on lower Mount Sharp.

Bottom: Curiosity's MAHLI camera captures a geometrically distinctive mineral formation at a mudstone outcrop at the base of Mount Sharp.

Above: Mudstone formations in
Death Valley National Park, California.

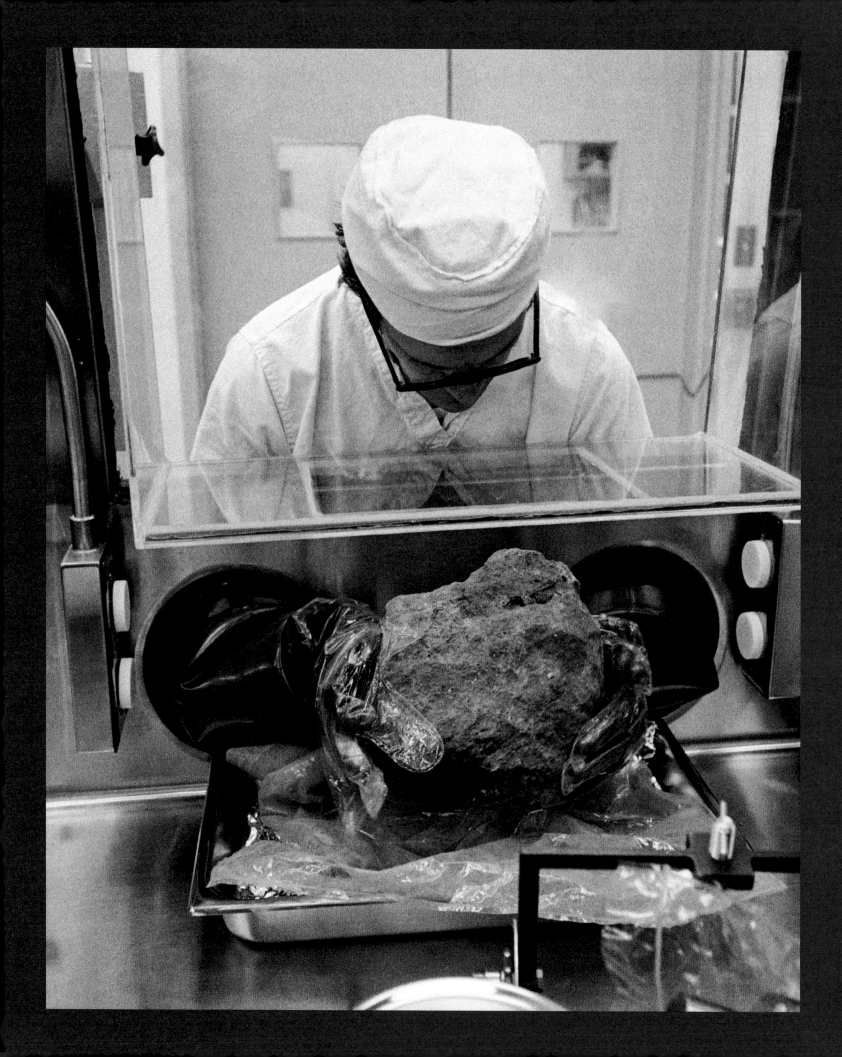

THE AGE OF PLANETARY SURFACES

Transcript of the Apollo 11 landing, from the spacelog

04 13 33 25 Aldrin: *Okay. Going to get the Contingency Sample there, Neil.*

04 13 33 27 Armstrong: *Right.*

04 13 33 58 Aldrin: *Okay. The Contingency Sample is down and it's up – [sampled?]. Looks like it's a little difficult to dig through the [surficial?] crust there.*

04 13 34 12 Armstrong: *This surface is very interesting. It's a very soft surface, but here and there where I plug with the Contingency Sample collector, I run into a very hard surface. But it appears to be very cohesive material of the same sort. I'll try to get a rock in here. Just a couple.*

04 13 34 54 Aldrin: *That looks beautiful from here, Neil.*

04 13 34 56 Armstrong: *It has a stark beauty all its own. It's like much of the high desert of the United States. It's different but it's very pretty out here. Be advised that a lot of the rock samples out here, the hard rock samples, have what appear to be vesicles in the surface. Also, I am looking at one now that appears to have some sort of phenocrysts.*

04 13 35 43 Aldrin: *Okay. The handle is off the [Contingency Sampler]. It pushes in about 6, 8 inches into the surface. [Looks] like it's pretty easy to [push in].*

04 13 35 56 Armstrong: *Yes, it is. I'm sure I could push it in farther, but it's hard for me to bend down further than that.*

04 13 37 08 Armstrong: *Contingency Sample is in the pocket.*

Earth before life, 4.5 billion years ago, was no nascent Eden; closer to the realm of Hades than the Garden of God. Molten and smouldering, cloaked in toxic clouds through which the young Sun could barely break to signal the coming dawn. The planet still trembled from the aftermath of a glancing collision with Mars-sized Theia, now fragmented into a ring of debris that, given time, coalesced with the ejected rubble of Earth to form the Moon. Four billion years later, two people who owed their existence to the energetic geology of Hadean Earth stepped onto the surface of the coalesced rubble and considered their position in the firmament. From their vantage point, Armstrong and Aldrin viewed a peaceful, pristine planet that has been many different worlds since its formation.

How do we know? How can we speak with any authority of events that happened billions of years ago, beyond not only memory but also life on Earth itself? How is such a timeline calibrated? The answer lies in the rocks.

Apollo delivered many treasures; countless engineering breakthroughs, a generation inspired, Earthrise from Apollo 8, the simple joy of exploration. But scientifically speaking, the treasure was rock; 382 kilograms of rock collected from six landing sites.

On the Moon, as on Earth, rocks can be dated with great accuracy using the natural clocks provided by the radioactive decay of certain atoms. The chemical element rubidium, for example, occurs naturally and quite commonly in a form known as rubidium-87, which is found in many potassium-rich minerals. It is unstable, with a half-life of 48 billion years. This means that over a period of 48 billion years, half of the rubidium-87 atoms that were present in a rock when it formed will have decayed away, transmuting into another sort of atom; strontium-87. An older rock will have fewer rubidium-87 atoms and more strontium-87 atoms. There is a little more to it than that, of course; how do we know how many rubidium and strontium atoms were present when the rock formed? The clever part of the dating procedure is that we don't need to know. The method, called the isochron method, relies on the fact that there is another naturally occurring form of strontium which is not produced by radioactive decay, known as strontium-86. Strontium-86 is stable and chemically identical to strontium-87; the only difference is that it has one more neutron inside its nucleus. This means that any strontium-86 atoms inside a rock today were present when the rock originally formed. By counting the number of rubidium-87, strontium-87 and strontium-86 atoms in a selection of samples taken from a rock, it is possible to calculate the absolute time elapsed since the rock formed.

On Earth, some of the oldest crustal rocks are found off the south-west coast of Greenland at a place called Isua. Using rubidium–strontium dating, we know these rocks were formed 3.66 billion years ago, with an uncertainty of 0.06 billion years. The oldest-known rock on Earth was found in Jack Hills, Western Australia. It formed 4.404 billion years ago, with an uncertainty of 0.008 billion years. All known samples from Earth's crust are younger than 4 billion years old, because the seething surface of the Hadean Earth meant that rocks were constantly being melted and reformed, resetting the radiometric clocks.

Opposite: Landing on the Moon with a manned spacecraft enabled scientists to get their hands on rocks for analysis.

'It has a stark beauty all its own.'
Neil Armstrong

The same analysis can be carried out on meteorites that have fallen to the surface of Earth from space. The majority (over a thousand known) formed between 4.4 and 4.6 billion years ago, which is consistent with independent estimates for the age of the Solar System, using, for example, helioseismology measurements of the amount of helium in the Sun's core.

The youngest of the Moon rocks returned by the Apollo astronauts were 3.2 billion years old, and the oldest were 4.5 billion years old. Twelve of the samples were over 4.2 billion years old. The variation in age of the Moon rocks at the Apollo landing site is extremely interesting and important, and it forms the basis of the most accurate technique for dating areas of the Moon from which we have no rock samples.

The graph opposite shows the age of the rocks collected from each Apollo landing site, and also the Russian Luna 24 robotic sample return mission, plotted against the number of craters per square kilometre at the site. The darkest blobs around the labels of the various missions represent the uncertainty of the measurements of the ages of the rocks and the crater counts. The two very low and uncertain points labelled Copernicus and Tycho are from rocks collected by Apollo that are thought to have been thrown into the landing sites by the impacts that created these two distinctive lunar craters in the much more recent past. Focus first on the two solid lines that curve up across the plot, bounding the measured points. The older sites

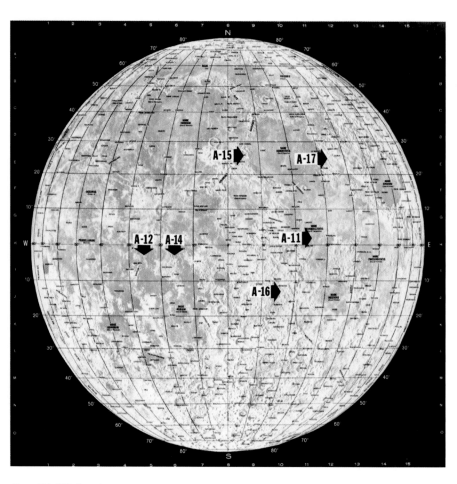

Above: Astronauts with soil-collecting devices, training for Apollo 16 mission (top) and on the Moon on Apollo 12 (bottom). Both missions successfully gathered rock and soil samples.

Above: This 1972 chart shows Apollo missions 11, 12, 14 and 15 at their respective landing points. The target locations for the future Apollo 16 and Apollo 17 missions appear here, too.

APOLLO LANDING SITES AND MASS OF ROCK SAMPLES RETURNED

Mission	Arrival date	Landing site	Latitude	Longitude	Sample
Apollo 11	20 Jul 1969	Mare Tranquillitatis	0°67'N	23°49'E	21.6 kg
Apollo 12	19 Nov 1969	Oceanus Procellarum	3°12'S	23°23'W	34.3 kg
Apollo 12	31 Jan 1971	Fra Mauro	3°40'S	17°28'E	42.6 kg
Apollo 12	30 Jul 1971	Hadley-Apennine	26°6'N	3°39'E	77.3 kg
Apollo 12	21 Apr 1972	Descartes	9°00'N	15°31'E	95.7 kg
Apollo 12	11 Dec 1972	Taurus-Littrow	20°10'N	30°46'E	110.5 kg
Luna 16	20 Sep 1970	Mare Fecunditatis	0°41'S	56°18'E	100 g
Luna 20	21 Feb 1972	Apollonius Highlands	3°32'S	56°33'E	30 g
Luna 24	18 Aug 1976	Mare Crisium	12°45'N	60°12'E	170 g

ANALYSIS OF CRATER SIZE AND THE AGE OF ROCK SAMPLES TAKEN

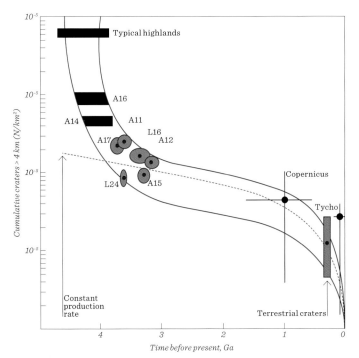

WELL-DATED SURFACES

AVERAGE VALUES   **A** APOLLO   **L** LUNA

SURFACE AGE NOT WELL KNOWN

have more craters, which is easy to understand, because there has been more time for meteorite impacts to build up. Surfaces that were created in the more recent past, such as the Copernicus crater, have experienced many fewer subsequent impacts.

Imagine now that there is an area of the Moon from where we have no samples, but we do have photographs of that location from space. We count the number of craters per square kilometre – let's say we get an average of 0.0002 craters above a diameter of 4 kilometres per square kilometre. This allows us to say that the surface formed somewhere between around 3.2 and 4 billion years ago, similar in age to the Apollo 17 landing site. This is one of the simplest uses of the graph.

More importantly for our purposes, we can extrapolate the graph from the Moon to Mars. We have to make some model-dependent estimates of how the number of impacts changes because Mars is a larger planet with a stronger gravitational pull, and to account for Mars's different place in the Solar System closer to the asteroid belt, the source of most of the impacting objects. We can do this with some confidence, and this is the primary way we estimate the absolute age of the different regions of Mars. The Noachian terrain is the oldest, and has the highest density of craters, followed by the Hesperian and the Amazonian. The Northern Lowlands have been resurfaced by volcanic activity in the more recent past, which we know because they are relatively devoid of craters, much like the lunar seas. In this way, the absolute dates we quote for events on Mars are all ultimately tied to the radioactive dating of Moon rocks, and this is one of the main reasons why the Apollo rock samples from different landing sites are so valuable scientifically.

There is another key line on the graph – the dashed line labelled 'Constant production rate'. The number of craters plotted against age would follow this line if the Moon had been subjected to a constant rate of impacts throughout its history. The measurements follow this line until we get to the surfaces around 3.8 to 4 billion years old, when the number of craters increases dramatically, implying that in the early history of the Solar System there was a time when the rate of impacts was much greater than today. That's not surprising – we might expect that the young Solar System would have been filled with leftover rubble from the formation of planets, and so a more violent place. But there is a complication. If we assume that the impact rates seen on the oldest lunar terrains, such as the Apollo 16 landing site in the Descartes Highlands, were sustained all the way back to the Moon's formation, then the impact rate would have been impossibly high; the amount of mass falling on the surface of the Moon would have been similar to the mass of the Moon itself! We know that didn't happen, so the assumption is that the rise of the impact rate curve on the graph is in fact a spike that peaked around 3.9 billion years ago, and then returned to the much lower impact rate from earlier times. This violent spike in the rate of impacts from space is known as the Late Heavy Bombardment.

The cause of the Late Heavy Bombardment is not known, but a leading theory is that Neptune changed its orbit from inside to outside that of Uranus, and the resulting gravitational disturbance deflected a maelstrom of icy objects in the distant Kuiper Belt towards the planets of the inner Solar System. We might reflect on the interconnected nature of knowledge and of the Solar System; exploration of the Moon provided us with the intellectual tools to date the surface of Mars, using the scars delivered by objects deflected into the inner Solar System by events that occurred far away in space and time, beyond the orbit of either world.

Top: Just a few examples of the mass and locations of soil and rock samples that were collected from the Moon between 1969 and 1976.

Bottom: This chart shows the number of large craters per km² at different Apollo and Luna (Russian lunar landing) sites, plotted against the age of the rock samples collected from the sites.

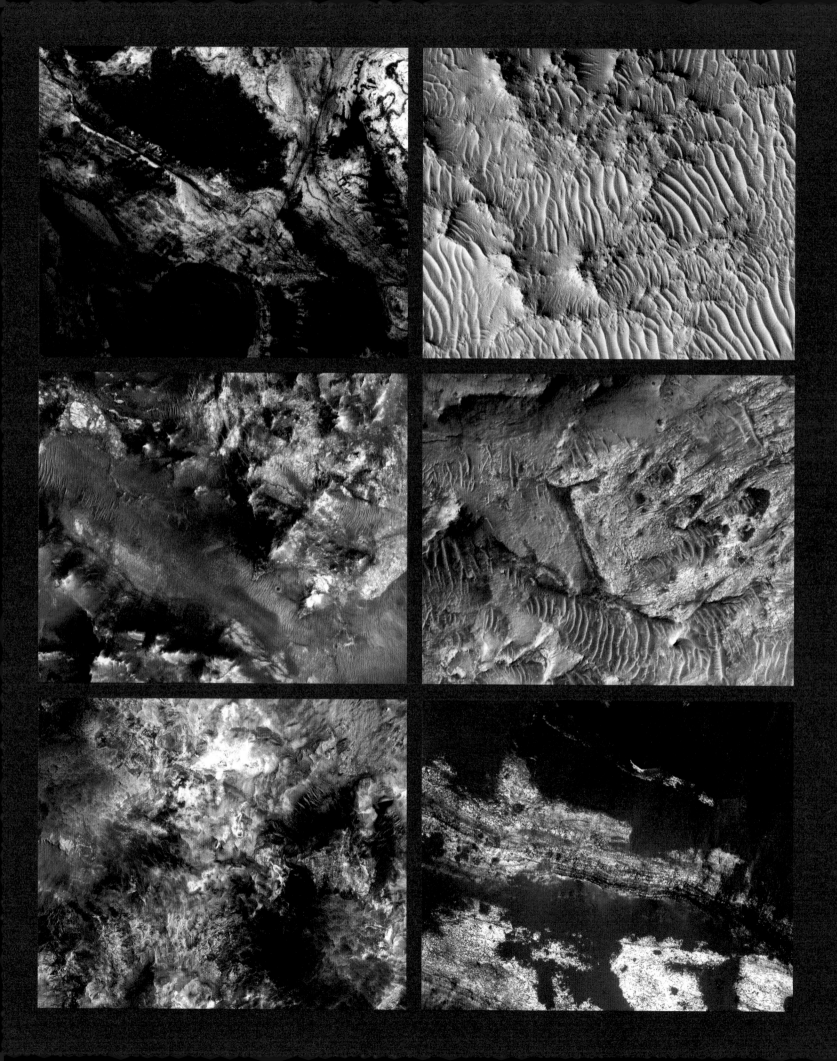

WRITING THE HISTORY OF MARS

Above: The Jezero crater region bears the marks of ancient rivers, and its soil contains carbonates and clays (green), indicating the presence of liquid water.

The crater-counting techniques developed from our exploration of the Moon allow us to place our observations of Mars today in historical context; they allow us to construct an absolute timeline. If we see evidence that water once flowed across a surface or pooled within a crater, and we can date the area using crater counting, then we can estimate when the water flowed, and when it disappeared. As we've discussed, Curiosity is exploring Gale and painting a vivid picture of what happened there. Crater counting allows for an estimate of when, independent from the rover's ground-based measurements. A recent analysis using images from orbit taken by MRO found 375 craters ranging in size from 88 metres to 23 kilometres inside Gale crater, implying an age of 3.61 billion years, with an uncertainty of around 60 million years. This places the formation of the crater and the geological events that played out inside it during the late Noachian/early Hesperian, and this timeframe is consistent with observations of the presence of water at many other sites of a similar age. Consistency over a wide range of independent observations is extremely important in science. One site or set of observations alone may be tantalising, but rarely compelling. Multiple independent observations reduce the uncertainty and give scientists more confidence in their understanding. When our goal is to understand the sequence of events on an ancient landscape far away that no human has ever visited, caution and consistency are both vital commodities.

Since we only have one active rover at present and landings are by their nature rare and confined to limited geographical regions, orbital observations of much wider areas of the planet are a vital tool. MRO's HiRISE camera has provided some of the most beautiful and detailed high-resolution photographs in the history of space exploration, covering over 99 per cent of the Martian surface. Features as small as a metre wide can be seen in these images, and the camera can also operate in the infrared, allowing for the identification of different minerals at the surface. There are many beautiful photographs in the MRO library that reveal evidence of persistent lakes and river systems, particularly across the Noachian terrain in the Southern Highlands. Sites such as the Jezero crater clearly show meandering river channels draining into deltas. The image shown in the picture left, constructed using both visible light and data from the spacecraft's spectrometer, reveals well-defined river channels flowing out of the crater and the presence of carbonates and clays that form where there is persistent liquid water. For these reasons, the Jezero crater is one of the three prime targets for NASA's Mars 2020 rover.

Some of the more recent finds are associated with the Arabia Terra region, which sits between the Noachian Southern Highlands and the great lowlands of the North, where water could have flowed towards a possible northern ocean. Researchers from University College London and the Open University studied an area of Arabia Terra equivalent to the size of Brazil using high-resolution imagery from MRO. The study mapped a network of over 17,000 kilometres of ancient river channels; not the usual excavated dry river valleys but rather a network of inverted channels – raised lines of sand and gravel laid down on the river beds that persisted while the land around them eroded away long after the rivers dried. Similar 'ghost rivers' are seen in desert environments on Earth in Utah, Oman and Egypt, where erosion rates are slow. The identification of the ghost rivers allows for Arabia Terra to be identified as a fossilised floodplain somewhat like the lower sections of the River Ganges on Earth – the lowland interface between mountain rivers and the sea.

'The weight of geochemical and geomorphological evidence points towards a late Noachian hydrological cycle that was intermittent, not permanently active.'
Robin Wordsworth, planetary scientist

This is an important discovery, because it feeds into a long-standing and still unresolved scientific debate about the nature of the Noachian climate. Nobody disputes that rivers once flowed on Mars, or that many impact craters were once home to lakes. The key question is one of longevity. Was Mars warm and wet for long periods of time, or was it for the most part a frozen planet on which water occasionally melted, perhaps triggered by sporadic volcanic activity or shifts in the planet's orbit? The tension arises because climate modellers find it very difficult to simulate an ancient Martian atmosphere thick enough to deliver a stable climate over hundreds of millions of years that would have been lost at a sufficiently fast rate to produce the Mars we see today. In a recent review, for example, Robin Wordsworth concludes, 'The weight of geochemical and geomorphological evidence points towards a late Noachian hydrological cycle that was intermittent, not permanently active.' And the latest state-of-the-art Martian climate simulations '... suggest a water-limited early Mars with episodic melting episodes may be a suitable paradigm for much of the late Noachian and early Hesperian climate.'

Before the recent analysis of the Mars Global Surveyor data, it was thought that there was a relative absence of river channels on Arabia Terra, and this was interpreted as support for a frozen, episodically wet Mars. Ice may have been

Top: In the Aeolis region, wind erosion has exposed and inverted a plethora of ancient channels – stream beds – in a fan-shaped sedimentary rock unit.

Bottom: We have similar inverted river channels here on Earth, such as these in Utah.

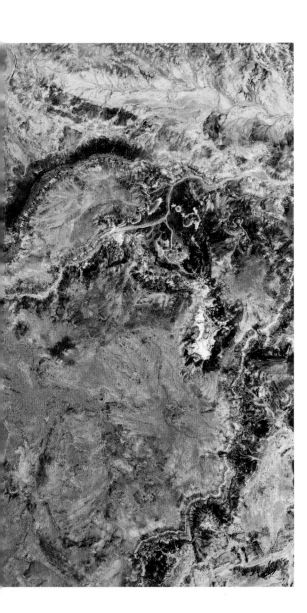

concentrated in the Southern Highlands and so lower elevations like Arabia Terra may have been generally more arid and therefore devoid of river channels. This appears not to be the case; Arabia Terra was once very wet indeed.

This debate has no consensus resolution at the time of writing, which reminds us of important points about both our knowledge of Mars and science in general. We are describing and attempting to understand and contextualise events that happened over 3.5 billion years ago, at or around the time that life on Earth began. That is an immense span of time, during which orbital parameters have changed, solar output has varied and innumerable impact events have cratered the planets. Tracing the evolution of an atmosphere backwards through time from today's measurements, anchored by geological and geochemical evidence from a few landing sites and orbital imagery, is extremely difficult. Suffice to say, there remains an unresolved tension between Mars's small size and large distance from the Sun and the clear evidence of an ancient water cycle, episodic or persistent. As Wordsworth writes in the introduction to his review, 'The nature of the early Martian climate is one of the major unanswered questions of planetary science.'

The complexity and apparent intractability of the mystery is reason enough for many scientists to study Mars. The character of the scientist requires not only comfort with but attraction to the unknown; an acceptance of and delight in the complexity of Nature. There is no simple story of any planet; they are each too big and too old and subject to too many variables. We might reflect on what that means for our own existence, as individuals and as a species. The specific fact of our existence is inextricably linked with the evolution of our planet through an incomprehensible web of cause and effect stretching back via major and minor events, many beyond Earth itself, and filtered by the multi-dimensional sieve of evolution by natural selection. This connects us with every living thing that went before, back to the origin of life almost 4 billion years ago and the origin of our planet half a billion years before that, and the origin of the Solar System and the laws of nature who knows when – perhaps during the Big Bang, or perhaps infinitely far away in time. There's a lot we can't know; the search for certainty is a fool's errand, and the lesson is to find delight in not knowing while simultaneously committing to extending the domain of the known. That's the key to science, the key to happiness and the only reasonable response to the existential challenge of existence, because it is the truth.

If you don't like that, comfort yourself with the fact that the domain of the knowable is vast, possibly infinite, and perhaps paradoxically the knowable things about Mars may exceed those about Earth. Because Mars has been in deep geological freeze for much of its life, it is easier to research the deep past than it is on Earth. And that deep past may include the origin of life. Traces of the geochemical alchemy that led to the emergence of self-replicating, information-carrying carbon molecules on Earth 4 billion years ago weathered away long ago, although clues may be found in the common structures and chemistry of living things today. On Mars, however, geochemistry was frozen and archived in regions such as Arabia Terra. If, therefore, life began or even still exists subsurface on Mars, such a discovery may deliver insight into our own origins that would simply not be possible by studying the weathered, ever-changing ecosystems of Earth. This may be the gift of unknowable Mars; she's less unknowable than Earth.

THE ERIDANIA SEA

The Eridania region of the Southern Highlands may provide the most compelling evidence yet of a warm, wet, early Mars with conditions not dissimilar to Earth at the time when life began. The analysis we describe below is a beautiful example of how the suite of science instruments aboard multiple spacecraft in orbit around Mars can be combined with our knowledge of cratering rates on the Moon to characterise an environment and trace its history back through deep time. It is also a beautiful example of the interconnected nature of science and the value of exploration.

We've already met the elevation maps from the MOLA aboard the Mars Global Surveyor. This topological data, in concert with visual data from MRO revealing the location of valley networks, suggests that the Eridania Basin was once a giant lake – possibly the largest ever to exist on Mars – 1.5 kilometres deep in places and containing three times more water by volume than the Caspian sea on Earth. We know the terrain is late-Noachian– around 3.7 billion years old – from crater counting. The CRISM spectrometer onboard MRO reveals the mineralogical composition of different areas of the Eridania Basin, and this can be used in conjunction with the topological and visual data to build a three-dimensional picture of the landscape and its chemical composition.

In the deepest areas of the Eridania Basin, magnesium- and iron-rich clay minerals are common, laid down as sheets of saponite, talc, serpentine and so-called TOT clays. These are characteristic of seafloor environments on Earth. Materials known as iron-rich phyllosilicates are also seen, which are again characteristic of terrestrial seafloor deposits. The mineral jarosite is detected, suggestive of the chemical weathering of sulphite deposits. There are many different carbonate signatures, containing iron, manganese, magnesium and calcium. Higher up the basin walls, chlorides are found, which are indicative of evaporation in shallower water. The most likely explanation for the origin of the deep clay, carbonate and sulphite deposits, it is argued, is volcanic activity on the floor of a deep lake, because very similar deposits are seen at hydrothermal sites on Earth. The iron minerals detected are similar to those formed in the iron-rich early oceans of Earth prior to the release of large amounts of oxygen into our atmosphere by photosynthesis.

All this beautiful, topological, visual, geophysical and geochemical data can be combined to produce a picture of what was happening in the Eridania Basin around 3.7 billion years ago. The geochemistry of the low-lying regions is indicative of a deep-sea, volcanically active, hydrothermal environment, with the salt-rich deposits at higher elevations suggestive of slow evaporation in shallower waters. The basin contained an iron- and mineral-rich ocean, energised by active volcanism deep below the waves. A cauldron of complex geochemistry stirred by the constant flow of energy from the inner planet. This is a striking conclusion, because at the same time a few hundred million kilometres across the Solar System on Earth, many biologists believe that life began in similar deep-sea volcanic environments.

The theory that life on Earth began in hydrothermal vent systems is mainstream, although not universally accepted by any means. The idea is partly based on the observation that all living organisms on Earth today share a chemistry associated with the establishment of gradients in the concentration of protons across membranes (known as chemiosmosis). Proton gradients are also characteristic of the geochemistry of alkaline hydrothermal vent systems in acid oceans which were common on the young Earth and, perhaps, the young Mars. An analysis published

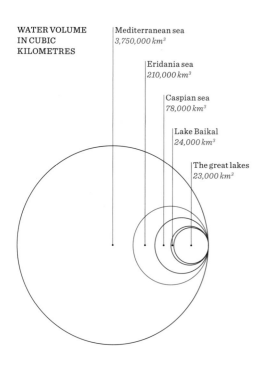

WATER VOLUME IN CUBIC KILOMETRES

Mediterranean sea
3,750,000 km³

Eridania sea
210,000 km³

Caspian sea
78,000 km³

Lake Baikal
24,000 km³

The great lakes
23,000 km³

Above: The Eridania Basin on Mars was once a sea or a great lake. Here the volume of water it likely held is shown in comparison to similar bodies of water on Earth.

'We shall not cease from exploration,
And the end of all our exploring
Will be to arrive where we started
And know the place for the first time.'
T. S. Eliot

ESTIMATED WATER DEPTHS ON MARS

estimated water depth (m)

>1000   700   400   100

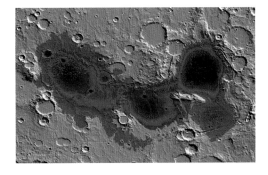

Top: The Eridania Basin of southern Mars is believed to have held a sea about 3.7 billion years ago, with seafloor deposits likely resulting from underwater hydrothermal activity. This graphic shows estimated depths of water in that ancient sea.

Bottom: Yellow iron oxide sediments on rocks near Tungurahua Volcano in Ecuador, on Earth. Similar mineral deposits have been detected in the Eridania Basin on Mars.

in 2017 claims to have detected evidence of biological activity in rocks laid down by hydrothermal activity at least 3.77 billion years ago in the Nuvvuagittuq crustal belt in Quebec (near the Greenland Isua rocks). The evidence comes from tube-like coils of the iron-oxide mineral haematite, similar in size and form to structures laid down by iron-oxidising bacteria in hydrothermal environments today.

It is possible that the rocks at Nuvvuagittuq are much older – a radiometric date using the samarium-neodymium decay process suggests an age of 4.28 billion years, which would make them by far the oldest rocks on Earth and an astonishingly early date for biological activity. These findings are controversial, in part because of the difficulty of understanding how Earth's active geology has modified the samples over such vast timescales. This is where Mars may have an advantage over Earth in helping us to understand the processes by which the chemistry of a planet became the chemistry of life. The environment at Eridania is far better preserved than that at Nuvvuagittuq. It has been in sterile, almost deep-freeze conditions for over 3 billion years; the ultimate laboratory for studying the origin of life, if indeed life began on Mars.

The story of the Eridania Basin and the possible scientific promise it holds was pieced together by using the results from different instruments on different spacecraft over many years, spanning several scientific disciplines: geology, chemistry, spectroscopy, laser altitude ranging and photography. The estimate of the age of the surface required the Apollo lunar rock samples from 50 years ago, and radiometric dating techniques which require an understanding of nuclear physics. The estimate of the age of the surface requires a model of the entire Solar System in order to interpret the measured crater density, which illustrates another important idea. The Solar System is a system; no planet is an island; no planet can be understood in isolation, just as the structure of any one living thing on Earth cannot be understood in isolation. Organisms are a product of evolution by natural selection, the interaction of the expression of genetic mutations and mixing with other organisms, in the ecosystem and the wider environment. The planets formed in a chaotic maelstrom from motions as random as the impact of a cosmic ray on a strand of primordial DNA, and whatever worlds emerged from the chaos have had their histories shaped profoundly by their mutual interactions throughout their evolution; the Late Heavy Bombardment is a beautiful example.

If indeed we discover that life began in the Eridania Basin, what a magnificent illustration of the unity of knowledge that would be. Understanding developed over centuries, the life's work of thousands of explorers, field geologists, chemists, biologists, astronauts, engineers, distilled into the discovery of a second genesis. But more than that; because of the pristine nature of the ancient environments on Mars, such a discovery would deliver an insight into our own origins that may not be available to us here on Earth. Mars is a time capsule, containing the frozen chemistry of an ancient world similar to the Earth when life began. But while Earth's deep past has been largely erased by 4 billion years of geology and the action of a complex global living ecosystem which has transformed the planet and its atmosphere beyond all recognition, Mars is a fossil from the deep past. This raises the wonderful possibility that another world, albeit an integral part of our system, may be the key to decoding the mystery of the origin of life here; the reason we exist. It's almost as if we have to leave the cradle in order to truly understand it.

THE DEATH OF MARS

'Olympus Mons makes Mount Everest look like a sand castle. Olympus is so tall that it reaches beyond the bulk of the atmosphere.'
Peter Cawdron, Retrograde

Around 3.5 billion years ago, change was under way on Mars. The warmer, wetter episodes that characterised the planet for millions of years during the Noachian became less frequent and the climate drifted, imperceptibly at first, towards the colder, arid state we recognise today. The water that flowed freely over the surface during the Noachian became locked away in giant reservoirs of ice. Lakes turned solid and rivers ran dry. The planet wasn't dying; far from it. The fireworks of the Late Heavy Bombardment had faded, but rumbles from the deep maintained the violence. The shield volcanoes of Tharsis, including the mighty Olympus Mons, continued to build even as global temperatures plummeted and subsurface lava flows occasionally melted the reservoirs of ice, leading to catastrophic floods that challenge the imagination in their scale and violence. The age of Noah became the Hesperian; the age of floods; or rather less portentously and more accurately, the age of occasional floods.

The Hesperian Period is named after the Hesperia Planum, a giant lava field to the north-east of the Hellas Basin in the Southern Highlands. The plane was resurfaced partly by the volcano at its centre, Tyrrhenus Mons, and spans almost 2 million square kilometres. First photographed by Mariner 9 in 1972, it is a relatively flat landscape with few visible rock formations. The most prominent surface features are the craters that allow us to date it. Tyrrhenus Mons is a very different volcano to Olympus Mons and the giants of Tharsis. It is much more ancient, fully formed not long after the Hellas impact 4 billion years ago, and it erupted through water or frozen ground, a process that produces large amounts of ash rather than lava – what geologists call a pyroclastic eruption. As a result of its age and composition, Tyrrhenus was more prone to erosion than the giants of Tharsis, and today stands only 1.8 kilometres above Hesperia Planum.

The Hesperian Period was coincident with increased volcanic activity, particularly at Tharsis, but there is not necessarily a causal link between volcanic activity and the transition from the Noachian to the Hesperian. What is clear is that volcanism became the dominant geological force on Mars as the long-lived water cycles waned and cratering rates declined after the Late Heavy Bombardment around 3.5 billion years ago. This is seen in both surface features and mineralogy – Hesperian surfaces are richer in sulphates rather than the clays and carbonates of the Noachian. Sulphates form in the more acidic atmospheric conditions created by the release of sulphur dioxide from volcanoes.

While volcanic activity was the dominant force sculpting the smoother Hesperian landscapes, there are spectacular and notable exceptions where water reasserted its dominance, if only for a moment. Echus Chasma is an unremarkable valley today. Images taken by MRO reveal a deep, steep-sided depression lying to the north of the Mariner Valley. At 100 kilometres long, 10 kilometres wide and between 1 and 4 kilometres deep, it would be impressive on Earth but is dwarfed by the magnificent neighbouring valley system. Three and a half billion years ago on frozen Hesperian Mars, however, things were different.

Above: An orbital view of Olympus Mons, the tallest volcano in the Solar System.

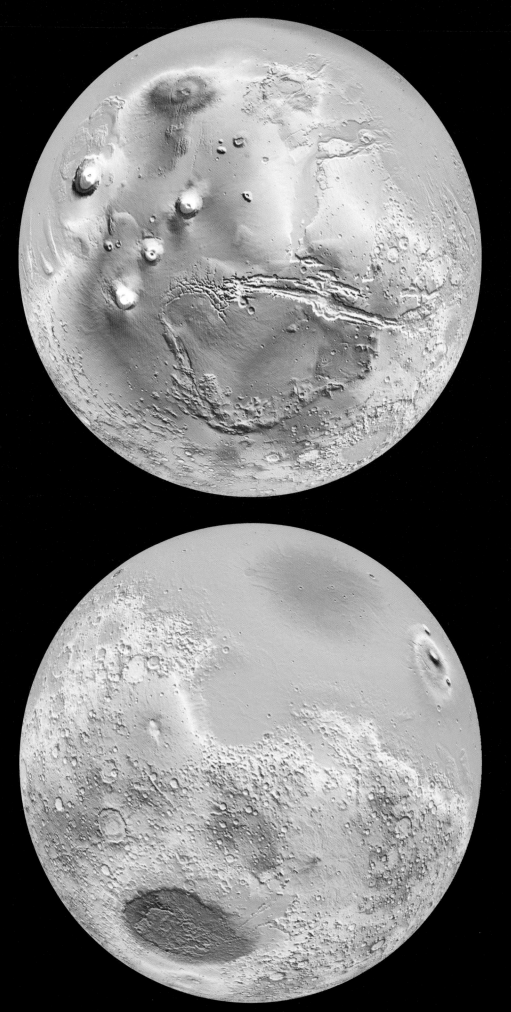

Global topographic views of Mars from the Mars Orbiter Laser Altimeter (MOLA). The bottom view features the Hellas impact basin (in purple, with red annulus of high standing material). The top view features the Tharsis topographic rise (in red and white). White and red represent the highest elevations, yellow indicates the equivalent of sea level, and the darker colours green, blue and purple indicate depths below.

'When we map out all the evidence of ancient liquid water on Mars, whether it's the dry river beds or the minerals that form in water, it really is restricted to the first billion years of its history. After that conditions quickly dry up and you end up with the Mars you have today, very cold, very dry and inhospitable for life.'
Ashwin Vasavada, planetary scientist

High above on the Tharsis Rise, lava surged upwards through the crust and met great reservoirs of ancient ice, releasing vast amounts of water that raged down from the Southern Highlands. The meltwater was funnelled towards Echus Chasma, where it created one of the briefest and yet most spectacular wonders of the Solar System.

The waters from the highlands came tumbling over the cliffs, roaring 4 kilometres down into the valley below over the largest-known waterfall in the history of the Solar System. Around 350,000 cubic kilometres of water cascaded through the valley – the equivalent of a cube of 70 x 70 x 70 kilometres – in less than two weeks. This being Hesperian Mars, the atmosphere was not substantial enough for the liquid to persist on the surface, and the water disappeared as quickly as it came, leaving scant evidence of the great falls etched into the canyon walls.

Echus Chasma was unique in scale but the cycle of volcanic eruption followed by fleeting catastrophic floods was repeated in many locations across the planet throughout the Hesperian. However, in the thinning atmosphere, the presence of

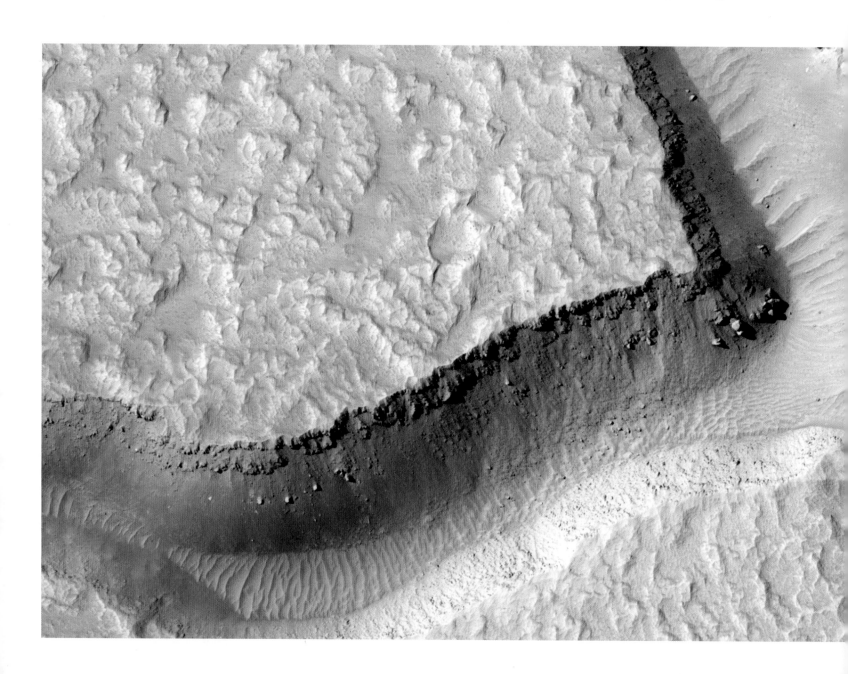

Opposite: Echus Chasma is thought to be the water-source region that formed Kasei Valles, a large valley that extends thousands of kilometres to the north.

Below: The ancient volcano Tyrrhenus Mons is made of layers that include volcanic ash rather than lava, producing broad channels of deposits.

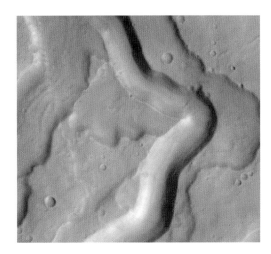

liquid water was probably always fleeting, increasingly so as time passed. Hesperian became Amazonian, around 3 billion years ago, plus or minus half a billion years or so. When a Mars expert is asked the question, 'When did the Amazonian begin?' they will usually say, 'Depends who you ask …' But certainly by 2 billion years ago, long before there was complex multicellular life on Earth, Mars was pretty much indistinguishable from the planet we see today.

The red planet didn't die entirely; volcanic eruptions persisted and have occurred in the very recent past at Tharsis; there may yet be eruptions in the future. But we do know that Mars changed from a world that supported liquid water on the surface under a thick atmosphere, via a period dominated by intense volcanic activity and episodic floods, into a hyper-arid frozen planet with a tenuous atmosphere more befitting its small size and place in the Solar System. This drama played out, very roughly, during the first 1.5 billion years of its life. The challenge is to understand why.

Right: This colourful scene is situated in the Noctis Labyrinthus, perched high on the Tharsis Rise in the upper reaches of the Valles Marineris canyon system. The dark sand dunes are made up of grains of iron-rich minerals from volcanic rock.

LOSING ATMOSPHERE

'For the first time in my life, I saw the horizon as a curved line. It was accentuated by a thin seam of dark blue light: the atmosphere. Obviously, this was not the "ocean" of air I had been told it was so many times in my life. I was terrified by its fragile appearance.'
Astronaut Ulf Merbold

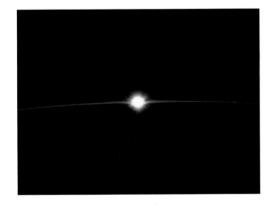

The climate of a rocky planet is a complex dynamic system shaped by myriad interactions, feedbacks and instabilities and inevitably changes over geological timescales. The thin envelope of atmospheric gases can be affected dramatically by volcanic and tectonic activity and impacts from space, and on longer timescales by shifts in orbit and tilt and changes in solar output. Earth seems to be a remarkable exception; the most dramatic atmospheric change over the last 3 billion years has been the gentle introduction of oxygen by photosynthesis.

Mars today has a thin atmosphere composed primarily of carbon dioxide. We know this atmosphere must have been significantly thicker in the past – quite possibly at an even higher pressure than that of modern-day Earth – because there was once liquid water on the surface. Mars must therefore have lost most of its original atmosphere, and it is important to understand where it went and why. The ancient atmospheric carbon dioxide molecules could have frozen out onto the surface in the polar ice caps, become locked away in the subsurface as carbonate minerals, or they could have been lost to space – or some combination of all three.

To measure the composition of the Martian atmosphere today, and understand how it evolved over time, a dedicated mission arrived in orbit in September 2014. MAVEN (the Mars Atmosphere and Volatile EvolutioN mission) has no high-definition cameras and provides no dramatic photographs of the surface, because the questions MAVEN was designed to answer are about the invisible envelope of gases surrounding the planet. If Curiosity is the extension of our hands and MRO our eyes, MAVEN is our nose, sniffing the air for clues to the past. The spacecraft operates in a highly elliptical orbit, bringing it as close as 150 kilometres above the surface and out beyond 6,000 kilometres. This eccentric path means that on every orbit, MAVEN experiences a wide profile of the Martian upper atmosphere, allowing it to build a three-dimensional picture of the gases that still remain and how their distribution changes with altitude and time.

MAVEN's measurements show that Mars is losing atmospheric gases today at a rate of between 2 and 2 kilograms a second. The question is, how is the atmosphere being lost and where is it going? The MAVEN team carried out an ingenious analysis based on observations of the concentrations of two different isotopes of the noble gas argon; argon-36 and argon-38. Argon-38 is chemically identical to argon-36, but it is heavier because it has two additional neutrons inside its nucleus. The ratio of argon-38 to argon-36 is sensitive to the rate at which the atmosphere is being lost to space through a process known as sputtering. High-speed electrically charged particles from the solar wind hit atoms in the upper atmosphere and literally knock them out into space. Because argon-36 is lighter than argon-38, the relative abundance of argon-36 in the upper atmosphere is enhanced – in technical terminology, argon-36 has a larger scale-height than argon-38. This means that argon-36 is preferentially lost to space through sputtering, and the ratio of argon-38 to argon-36 increases. Since there is no other way that argon can be removed from the atmosphere – it doesn't react with anything and doesn't freeze at anywhere near the temperatures seen on Mars – the argon measurements provide a very clean indication of the way the Martian atmosphere is affected by the solar wind.

Opposite: NASA's Mars Atmosphere and Volatile EvolutioN (MAVEN) spacecraft launches from Cape Canaveral. MAVEN is the first spacecraft devoted to exploring and understanding the Martian upper atmosphere.

Above: NASA astronaut Scott Kelly posted this striking photograph of the Sun peeking over the thin blue line of Earth's atmosphere to Twitter from the International Space Station, on 21 September 2015.

ATOMIC CARBON

distance from Mars (R_{Mars})

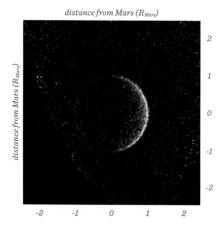

ATOMIC OXYGEN

distance from Mars (R_{Mars})

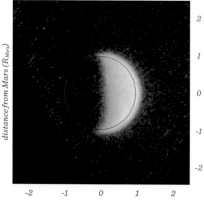

ATOMIC HYDROGEN

distance from Mars (R_{Mars})

By using measurements from MAVEN at different altitudes during its elliptical orbit, surface measurements of the argon isotope ratio from Curiosity, this beautifully elegant experiment enabled the team to determine that 66 per cent of the argon in the Martian atmosphere has been lost over the planet's 4.5-billion-year lifetime. Once this rate of argon loss has been determined, the rate of other atmospheric components can be calculated. The published results concluded:

The evidence from MAVEN observations suggests that a large fraction of the Martian volatile inventory has been removed to space and that loss to space has been an important process in the evolution of the Martian atmosphere through time. In particular, a large fraction of the dominant component of the Martian atmosphere, the powerful greenhouse gas carbon dioxide, has been lost to space through the interaction with the solar wind, with a smaller fraction being locked away in subsurface carbonate deposits and an even smaller fraction stored as frozen carbon dioxide in the polar ice caps. These changes, the authors conclude, appear to be large enough to account for the change in the Martian climate inferred from the planet's geomorphology.

Without declining levels of greenhouse gases in the atmosphere to keep in warmth and protect its lakes and oceans, Mars was destined to become a frozen desert world. But why has Earth not suffered the same fate? We are, after all, closer to the Sun and bathed in a more violent solar wind. Why has our atmosphere survived the solar onslaught for several billion years after the Martian atmosphere was stripped away?

Above: Three views of an escaping atmosphere, obtained by MAVEN's Imaging Ultraviolet Spectrograph. By observing all of the products of water and carbon dioxide breakdown, MAVEN's remote sensing team can characterise the processes that drive atmospheric loss on Mars. These processes may have transformed the planet from an early Earth-like climate to the cold and dry climate of today.

Above: At Mars's South Pole, the polar cap is made from solid carbon dioxide (dry ice), which does not occur naturally on Earth. New dry ice is constantly being added to this landscape by freezing directly out of the carbon dioxide atmosphere or falling as snow. At minus 130 degrees Celsius, this is the coldest landscape on either Earth or Mars.

Above: The Crab Nebula is one of the most intricately structured and highly dynamic objects ever observed. It is also the only known place in the Universe where noble gas molecules, in this case argon hydride, have been found occurring naturally.

EARTH'S PROTECTIVE SHIELD

Acentral idea in this book is that the Solar System is in fact a system; although we might feel isolated on a ball of rock in the dark, that is an illusion. No planet is an island. The system evolves as one. Draw a diagram of the food chain underpinning all complex life on Earth today, and at the base is a stream of photons generated 150 million kilometres away inside the Sun. Green plants are the interface between a star and civilisation.

Our understanding of Mars demonstrates that a planet's relationship with the Sun is not all light and sweetness. The star can provide the energy for life, but the solar wind can also destroy life or prevent it from ever emerging through the damage it can cause to planetary atmospheres.

The solar wind emerges from the Sun's atmosphere, known as the corona. The temperature at the Sun's surface is 6,000 degrees Celsius, but the temperature of the corona is over a million degrees Celsius, heated by energy transferred from the surface by the powerful solar magnetic field twisted and coiled by the star's rotation. At such high temperatures, atoms cannot hold together. Electrons are ripped away from atomic nuclei and matter exists in a fourth state; not solid, liquid or gas, but a soup of negatively charged electrons and positive ions known as a plasma. Some of these particles are moving so fast that they escape the Sun's embrace completely, streaming outwards across the Solar System at velocities in excess of 800 km/s. This is the solar wind.

109   Approximate
size of Earth

Top: A solar prominence
in full flow; the hot gas
dramatically bursts
outwards from the Sun.

EARTH + MARS

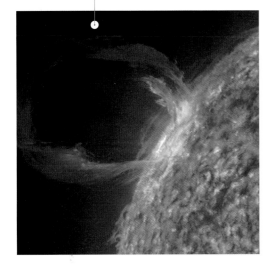

When these high-energy charged particles reach Earth, most do not smash directly into atoms at the top of our atmosphere because we are protected by our magnetic field. They are deflected harmlessly by stretching and distorting the field out on the night side of the planet. The stretched field occasionally reconfigures, accelerating electrically charged particles back down the magnetic field lines to the Earth's poles, where they collide with atoms and molecules in the upper atmosphere, exciting them and causing them to emit light. This is the aurora; the Northern and Southern lights.

From Earth the aurora can be seen by those lucky enough to be far enough north or south on the right night, when the sky is clear, the Sun is active and the interplanetary magnetic fields are favourably aligned. From the lofty vantage point of the International Space Station the aurora reveals itself as halos dancing around the poles. It is an unforgettable sight, undoubtedly one of Nature's wonders, but as is often the case in science, and in life for that matter, the experience is greatly enhanced by knowledge.

The lights dancing faster than the eye can see or perhaps the brain can comprehend are a direct result of the structure of atoms; electrons shifting between allowed energy levels determined by the number of protons in the atomic nucleus that holds them captive. High-altitude oxygen glows red. An atom is struck by a charged particle accelerated down the Earth's magnetic field lines towards the pole. The interaction shifts an electron into a long-lived excited configuration. If the atom is not involved in a collision in the thin air for around 100 seconds – an eternity in atomic time – the electron will shift down closer to the nucleus, and a red photon will be emitted. At lower altitudes, a different reconfiguration inside oxygen atoms causes the emission of green photons. This time the atom must avoid collision for around a second; still an atomic eternity in the denser air. Excited nitrogen molecules add a deep red or pink to the lower edges of the towering curtains of light during intense displays.

This is quantum mechanics written across the sky; the structure of atoms revealed. To calculate the colours, physicists treat the electrons as waves trapped inside a deep well created by the electric charge of the protons. Particles imagined not as tiny flecks of matter, but as fields that can span great volumes of space. The energy to drive the display comes from nuclear fusion reactions inside the Sun; a factory a million times the volume of Earth, converting hydrogen into helium at a rate of 600 million tonnes per second. The weak nuclear force acts slowly and grudgingly to convert protons into neutrons in the core – nuclear alchemy that results in the creation of neutrinos, which stream across 150 million kilometres at close to the speed of light before passing, unhindered, through the entire planet on their journey to infinity. Thus 60 billion per square centimetre stream through your head every second without even a nudge, because your head is almost entirely empty space and neutrinos hardly ever come close enough to interact with your sparsely distributed molecules. It might happen once in your lifetime.

The nuclear forces at work inside the Sun also assembled the heavy elements that make up your body inside old stars long ago from hydrogen and helium forged in the first few seconds after the Big Bang: carbon, oxygen, nitrogen, sulphur, phosphorus and iron. Gravity assembled the stars by forcing the collapse of primordial interstellar clouds, causing them to heat up in their centres and triggering the

Left: A 35 mm frame of the
Aurora Australis, also known
as the Southern Lights,
photographed from Space
Shuttle Discovery's flight deck.

Top: Our magnetic field is what protects Earth against the same fate as Mars. Without it the upper atmosphere would be stripped away.

EARTH'S MAGNETIC FIELD

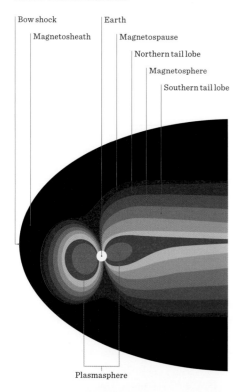

Bow shock

Magnetosheath

Earth

Magnetospause

Northern tail lobe

Magnetosphere

Southern tail lobe

Plasmasphere

nuclear reactions that release the energy to halt the collapse for a few billion years – long enough for life to emerge in the gradient between the nuclear furnaces and the chill of the expanding Universe. All of these thoughts are ignited in the imagination by the auroral display. Our experience is enhanced by knowledge.

A central component of the web of physical processes that make the aurora shine is our magnetic field, which deflects the solar wind harmlessly past our planet and allows it to return in a much-subdued form to the poles. Without this, the processes that MAVEN observes stripping the atmosphere of Mars would happen on Earth. The magnetic field is one of the most important differences between our worlds.

Earth's magnetic field is generated by what is known as a dynamo, which is easy to explain in broad-brush terms but extremely complex and not fully understood in detail. The basic physics is simple; Earth's core is composed largely of molten iron, which is an electrically conducting fluid. Plumes of molten liquid rise and cool as they approach the mantle, and the Earth's rotation swings these columns around into circulating rising and falling flows. This circulating flow acts as a dynamo and generates a magnetic field. The stability of the dynamo is related to the temperature gradient between the core and the mantle, the radius of the core itself, the rate of rotation of the planet, and numerous other subtleties. The rate of heat flow through the mantle is affected by geological processes including plate tectonics and volcanism. The precise nature of the flows depends on the constituents of the core other than iron; for example, the sulphur content and distribution. And so on. It's fiendishly complicated, and because of this complexity our planetary magnetic field gives every indication of being a delicate phenomenon. It was certainly delicate on Mars.

'As if from Heaven itself, great
curtains of delicate light hung and
trembled. Pale green and rose-pink,
and as transparent as the most
fragile fabric, and at the bottom
edge a profound and fiery crimson
like the fires of Hell, they swung
and shimmered loosely with more
grace than the most skillful dancer.'
The Subtle Knife, Philip Pullman

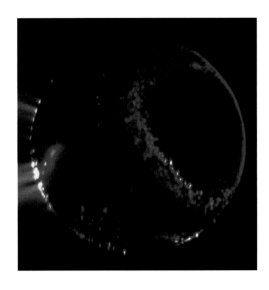

We suspect there was once a Martian dynamo because the Martian crust is
still partially magnetised in places, particularly in the ancient Southern Highlands,
suggesting that these rocks cooled in the presence of a global magnetic field. The
Hellas impact basin, which formed around 4 billion years ago, shows no sign of
magnetisation, suggesting the dynamo shut down before this impact. Impact basins
that formed over 4.05 billion years ago do show signs of magnetisation. The Martian
Meteorite ALH 84001, dated at 4.1 billion years old, also shows clear signs of having
formed in the presence of a magnetic field. These observations lead to the current
baseline view that Mars had a magnetic field not unlike Earth's, which seems to have
switched off just before the Hellas impact. From that point onwards, the atmosphere
was exposed to the full force of the solar wind, and the MAVEN results tell us that
this was likely the key driver for the loss of the Martian atmosphere.

It would be tempting, but wrong, to suggest that the loss of the magnetic field was
the trigger that drove the change from the Noachian to the Hesperian. The timescales
probably don't match, and in any case the loss of the magnetic field wouldn't have
led to an immediate loss of the atmosphere – sputtering is a gradual process which
still continues today. There have been speculative attempts to link the switch-off
of the dynamo to the Hellas impact or multiple impacts around the time of the Late
Heavy Bombardment. The idea is that the heating of the mantle can disrupt the
heat flow out from the core, in turn disrupting the smooth convective flows. At the
time of writing, the sequence of events is reasonably well established but there
is no consensus on how, or even if, they are causally linked. A complicating factor
is the relative instability of the Martian orbit. The planet's obliquity – the tilt of
its spin axis – varies by tens of degrees over timescales of a few hundred thousand
years in response to gravitational interactions with the other planets and the
Sun. In comparison, Earth's axis drifts gently between 22.1 and 24.5 degrees in
a 41,000-year cycle, and even this small change is responsible in part for Earth's
periodic ice ages. Imagine what happened to the Martian climate when the polar
regions with their reservoirs of carbon dioxide and water ice tilted towards the
Sun by 20 or 30 degrees. The stability of Earth's spin axis is maintained by the
damping effect of the Moon, which is unusually large relative to our planet; the
Earth–Moon system might almost be regarded as a double planet system. The more
we learn about other worlds, the more fortunate our presence on Earth appears.

Despite all this marvellous complexity, the simple thought with which we
began this chapter probably captures the spirit of why Mars died. The red planet
is simply too small. It lost its heat more quickly than Earth from its smaller core
and through its thinner mantle and crust, geological activity fell markedly after
reaching a peak during the Hesperian, and the planet faded. The atmosphere, held
more tenuously than Earth's because of the planet's lower mass and gravitational
pull and unprotected by a magnetic shield, was stripped away by the solar wind, and
the surface approached temperatures more in tune with an orbit far from the Sun.
Mars's fate was sealed during its formation 4.5 billion years ago; it is too small and
too far away from the Sun to have remained a vibrant, active world.

Today, faint auroras still dance in the Martian skies, as photographed by ESA's
Mars Express spacecraft; blue ghosts of a Martian past generated by localised
magnetic hotspots scattered across the planet and the last traces of carbon dioxide
in the atmosphere energised by the relentless solar winds.

Opposite: This illustration
depicts charged particles
from a solar storm stripping
away charged particles
of Mars's atmosphere, one
of the processes of Martian
atmosphere loss studied
by NASA's MAVEN mission.

Above: Scientists have
created this image of how
they predict Mars's auroras
might appear based on data
received from Mars Express.

Back on Earth, the protective shield of our atmosphere allows a dazzling array of ecosystems to flourish.

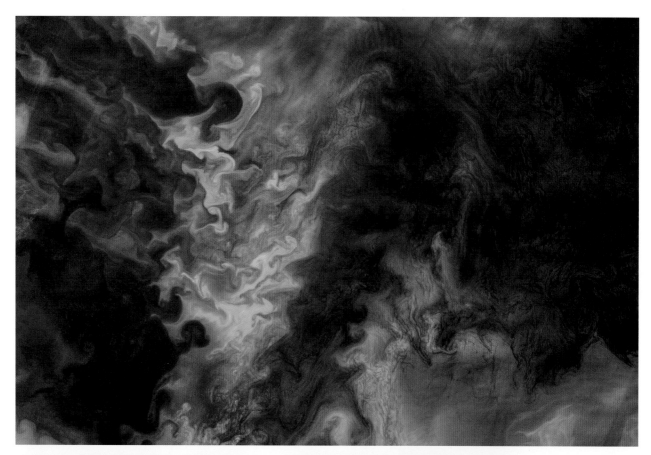

Right: Regardless of the amount of winter ice cover, the waters off the Alaskan coast come alive each spring with blooms of phytoplankton.

Right: In July 2018, an iceberg weighing 11 million tonnes parked just offshore of Innaarsuit, a small island village in Greenland.

Left: NASA astronaut Ricky Arnold captured this photograph of a changing landscape in the heart of Madagascar, as the International Space Station flew overhead.

Left: This night shot from the International Space Station taken 415 km above the English Channel shows the lights of the northern European cities.

VISIT THE
HISTORIC SITES
MARS
MULTIPLE TOURS AVAILABLE
ROBOTIC PIONEERS / ARTS & CULTURE / ARCHITECTURE & AGRICULTURE

THE FUTURE

'Do I think there's life today on Mars? If we found evidence it would obviously be a huge discovery. But I think it might actually be more surprising if we never found evidence of life on Mars. Everything we've found suggests that Mars was such a friendly, supportive place for life in its early history, and there should be a lot of planets like that around other stars and lots of life in the Universe.'

Ashwin Vasavada, planetary scientist

Opposite: This travel poster was created by the creative team at NASA's Jet Propulsion Lab to spark imaginations. Maybe one day Mars travel will be a reality!

Despite its 3-billion-year quiescence, Mars had a bright and active adolescence that echoes down the ages and may yet have profound consequences for us, confined for now on her planetary neighbour a solar system's lifetime away.

There may have been, and may still be, life on Mars. We're looking hard, and the next generation of spacecraft may find evidence for Martians within the next decade. The discovery of a second genesis would have important philosophical, scientific and cultural consequences. It would mean that, given the right conditions – liquid water, active geology and a sprinkling of organics – we would know that life emerges with a sense of inevitability through the entirely predictable action of the laws of Nature. It would be confirmation that we, along with every living thing on our planet today, are an outgrowth of the planet; an extension of geology. We would understand that whatever magic we perceive in the depths of our souls has its origins in the interface between water and heat and minerals and gas; we are bubbling chemistry that thinks. If it happened on two neighbouring planets in a single solar system, it happened everywhere. We would understand that we are part of a grand living universe; we are not God's children, but we are also not alone.

Of even greater importance, I think, is the role that Mars must play in our future. Mars is rich in potential. It has reservoirs of frozen or even liquid water below the surface and a great wealth of mineral resources; all the things that are necessary to support a civilisation. Because of its history, this is a world in waiting. A treasure island in stasis. I think in my lifetime there will be Martians. The Martians will be us. We will go to Mars and make it our home, partly because there is nowhere else to go. Mars is the only planet beyond Earth on which we could even contemplate a landing in the near future.

The message from Mariner 4 was not interpreted correctly. There is a second chance for us on Mars, when we feel ready to take it. I don't mean that we will leave Earth en masse for the new world; this is manifest nonsense. The Earth is by far and away the best planet we know of anywhere in the Universe; we were born of it and sculpted precisely and perfectly to flourish on it through the action of evolution by natural selection. Yet it is possible, now, to build a colony on Mars. I imagine a group of pioneers on the new world, living off the land and constructing the infrastructure necessary for tens, hundreds and ultimately thousands of individuals to follow in their footsteps across the Solar System and extend the human frontier for the first time in many centuries.

Frontiers are important. The intellectual frontier is the domain of science, and where would we be without the pioneers who felt drawn there? The physical frontier is the domain of adventurers, engineers and dreamers, and dreamers need somewhere to go. In Werner Herzog's film *Encounters at the End of the World*, a man named William Jirsa describes the sort of person who gravitates to Antarctica, the last of Earth's frontiers; 'I like to say, if you take everybody who's not tied down they all sort of fall down to the bottom of the planet, so that's how we got here, you know. We're all at loose ends, and here we are together. I remember when I first got down here I sort of enjoyed the sensation of recognizing people with my travel markings. I was like, hey, these are my people. PhDs washing dishes and linguists on a continent with no languages.'

Mars is the step beyond the Antarctic, and perhaps we all need to take it in our imaginations. Our civilisation feels untethered to me. We're at a loose end, huddled in a tiny corner of our system. We watch the wandering planets and the wheeling stars and the cycles of day and night and wonder what the hell we're doing; we're all linguists on a continent with no languages. As a consequence, we scrabble around, eking out the ever-more precious resources on Earth, trying to expand and build more stuff and grow in a thin shell of air on the two-dimensional surface of a small rock, giving little thought for the three-dimensional path marked by the lights in the night. Robert Zubrin, the visionary engineer whose work has inspired, amongst others, Elon Musk, likes to say that ideas have consequences, and the worst idea in the history of humanity is that we must compete for limited resources. This is false. The Solar System contains raw materials beyond our needs or desires, and they will become resources when we choose to access them. The international tensions created by the competition for Earth-bound resources are based on the entirely false and dangerous idea that resources are limited. False. False. False. We have the technology, and perhaps as we drift ever more aimlessly, we may discover the will, to unlock the unlimited treasures in the vast solar system of which we are a part.

We need to shift our collective consciousness. We burn too bright for this world alone, but that does not mean we should extinguish the flame. We must forge a new path that transcends competition between nations, that requires the wholesale rejection of the mindset that mistakenly compels us to retreat or fight over dwindling resources on a single groaning planet. We must transform ourselves into a multi-planetary spacefaring civilisation, and this begins with the colonisation of Mars; an achievable goal forged from the human desire to explore and expand without pushing up against someone else's border fence or damaging our planet beyond repair. Imagine the magnificent intellectual and physical vistas, the new technologies, the opportunities, the excitement of creating a new society, the joy of extending our collective experience and hopes and dreams to a new world, and worlds beyond.

Mars has a pivotal role to play in our future. If we don't go there, we'll never go anywhere, and if we go nowhere, we'll die here. Breathing new life into the old red planet will breathe new life into us; it will be our first step beyond the cradle and onwards to the stars.

Opposite: A stunning view of the Milky Way over the Southern Alps in New Zealand, with Mars putting in an appearance in the top left of the image.

JUPITER

THE GODFATHER

ANDREW COHEN

'Not all those who wander are lost.'
The Fellowship of the Ring,
J.R.R. Tolkien

'Jupiter is the King of the Planets,
not only in terms of its size and the
beauty you see in the cloud tops, but
also in terms of shaping the entire
evolutionary story of our Solar System.'
*Leigh Fletcher, scientist, Jupiter
Icy Moons Explorer mission*

ANCIENT GIANT

It's a story that began before the Solar System even truly existed. Five billion years ago, a vast interstellar cloud of dust and gas, at least 65 light years across (that's 6.1×10^{14} in kilometres), began to collapse and coalesce under its own enormous weight. Fragmenting into a collection of smaller, dense cores, one of these gigantic clumps formed what is known as the pre-solar nebula – the embryo from which every part of our Solar System would develop.

In this dense cloud of gas comprised almost entirely of hydrogen and helium hung every atom that would go on to form the Sun, the planets and every element that would define the character of each of these worlds – including the ingredients of life, the components of you and me.

What triggered this cloud beyond the tipping point and on to the next stage of its stellar evolution is lost in the aeons of time, but astonishingly we have been able to pull together scraps of evidence to guide us. By studying some of the oldest meteorites ever discovered on the surface of the Earth we have been able to get a glimpse of the time when it all began.

Locked away inside these ancient meteorites are the unmistakable signatures of rare forms of iron that we believe can only have been created under a very specific set of conditions. The only place in the Universe where these rare isotopes can be formed is at the heart of giant, short-lived stars at the very moment they explode – in other words, they are the product of supernovae.

This clear chemical fingerprint from rocks dating back to the earliest days of the Solar System, over 4.5 billion years ago, suggests that our Solar System emerged among a cluster of thousands of stars in a large stellar nursery, perhaps not unlike the Orion Nebula, the star-forming region that lies closest to the Earth, at 1,500 light years away, which can even be spotted with the naked eye in the night sky.

Amongst this original stellar cluster were massive stars that would have dwarfed our own, all burning hard and fast. It appears that as one of these stars reached the end of its life, its death created a supernova and with it a shockwave that rippled through the distant origins of our Solar System, causing dense regions within the pre-solar nebula to rapidly collapse. This moment provided the precise conditions for a new star to form, and thus the story of our solar system began.

For the first 50 million years after the Sun was born, our corner of the Universe was shrouded in darkness, the only light the dim red glow of our embryonic star – a T Tauri-type star – struggling to shine through the very earliest years of its infancy. From that original cloud of gas, the Sun forming at its centre would swallow up 99 per cent of all the matter in the Solar System, leaving just a tiny fraction that would go on to form the other planets.

Yet when, after around 50 million years, the Sun's nuclear furnace ignited, its light revealed that one giant planet had already scooped up much of the remaining matter; a world 300 times more massive than the Earth is today. Jupiter, the oldest of all the planets, was there to witness the Sun's first dawn. In fact, we think Jupiter had been there from almost the very beginning of the Solar System. Evidence suggests that just 1 million years after the birth of the Sun, Jupiter had already formed a core 20 times more massive than the Earth, orbiting the dark Sun and dividing the building material of the Solar System into two distinct regions, each with a slightly different chemical makeup, that we would be able to observe in the signature of meteorites that fell to Earth billions of years later.

Within just a few million years this young giant was 50 times as massive as the Earth, and it would go on growing in the dark. By the time the temperature and pressure at the centre of the Sun had risen so high that hydrogen began to fuse and light up the heavens (entering the main sequence phase of its life in which it continues today), Jupiter was fully grown. Under the cover of darkness before the Sun's first dawn, this planet had taken up its position of power, a formidable force that would transform the destiny of every planet that would follow.

Opposite: At only 1,500 light years away, the Orion Nebula is the closest star-forming region to Earth. It is from a stellar nursery like this that Earth was probably formed.

Top: This hot planet is four times the mass of Jupiter, but as a young planet it is still contracting under gravity and radiating heat.

Bottom: A young star, much like our own in its early history, in the centre of this nebula is beginning to contract to become a main sequence star.

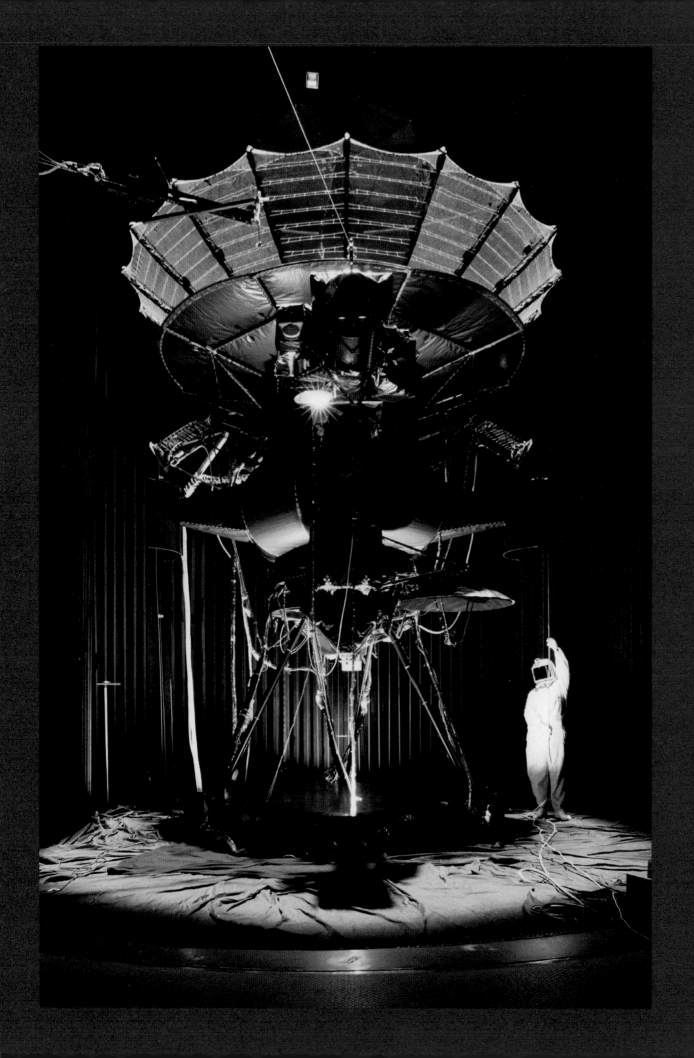

KING OF
PLANETS

'Jupiter is the biggest planet in this
whole system. It is the giant. Its
diameter is ten times the size of the
Earth. Its mass is three hundred and
twenty times the mass of the Earth.
This is a big giant planet. It is five
times the distance between the Earth
and the Sun, out there from the Sun.'
Fran Bagenal, New Horizons mission

Approximate
size of Earth

To the Romans, Jupiter was the king of the gods, the ruler of the heavens and
a figure of power who drove forward the all-conquering Roman army with a
thunderbolt in one hand and an eagle by his side. The name Jupiter itself means
skyfather, and it is constructed from the Latin words *dies* (day or sky) and *pater*
(father) and, as is often the case with mythology, there is a truth lurking amongst
the legend.

In size alone Jupiter is by far the biggest planet in the Solar System. Its mass
is a staggering 1,900 trillion trillion kilograms – that's two and a half times bigger
than all of the other planets in our Solar System combined. It is so big, in fact, that
if you were an alien species observing our system from a distant part of the galaxy,
the Sun and Jupiter might well be all you could see.

With a diameter of 142,984 kilometres at the Jovian equator, the Earth would
fit inside Jupiter over 1,300 times, and yet this giant of a planet is made almost
entirely of gas. Comprising 89 per cent hydrogen and 10 per cent helium, it is far
less condensed than any of the rocky planets and has a density far smaller than
that of the Earth – 1.326 g/cm³ compared with Earth's 5.513 g/cm³. For all of its
size, Jupiter also has the shortest day of any planet, completing a single rotation
every 9 hours 55 minutes and 30 seconds as it whips around on its axis at 45,000
kilometres an hour.

In the last 50 years our knowledge of Jupiter has grown exponentially, and with
it we have gained a deeper understanding of the role it has played in the shaping
of the Solar System. No longer just the beautiful banded world we see through our
telescopes, this is a planet that we have explored up close, our eyes and ears living
within the Jovian system for years at a time.

The first close-up glimpse of the giant planet came with the cameras of the
Pioneer 10 spacecraft in 1973. This was the first of our planetary explorers to cross
the great boundary of the asteroid belt, and from an approach of just over 130,000
kilometres it was able to take the first close-up images. In total, Pioneer 10 sent back
over 500 precious pictures of this distant world before flying on to the very edges of
the Solar System to complete its mission. Just over a year later, Pioneer 11 followed
hot on the heels of its sibling craft and swept within 43,000 kilometres of Jupiter's
cloud tops before heading on towards Saturn. Voyager 1 and 2 passed by Jupiter
in the late 1970s, taking the first extraordinary images of not just the planet but its
moons, too, as part of their own grand tour of the Solar System. This was an historic
moment, as Fran Bagenal, scientist on the New Horizons mission, explains, 'We
had hints, but it wasn't until Voyager went up close and personal and took amazing
pictures of the moons of the planet that we really began to get a sense of just how
complex the worlds around Jupiter are, and what a variety there is.'

The Galileo spacecraft became our first explorer to actually live in the Jovian
system. Entering orbit around the planet on 7 December 1995, Galileo spent two years
touring Jupiter, allowing us to explore as never before, including an audacious deep
dive directly into the planet with a probe that was released from the craft and that
collected data for almost an hour as it parachuted down through 150 kilometres of
Jupiter's atmosphere before being destroyed by the intense pressure and temperature.

Opposite: The Galileo Jupiter
spacecraft receives its last
tweaks and checks before
launch in October 1989.

Above: The vast size of Jupiter
can begin to be appreciated
when viewed in comparison
with Earth, marked here.

'The Galileo mission was the first in which we went into orbit around Jupiter and were able to send a probe into the atmosphere. We couldn't find any water. We didn't find any clouds. The big question after the Galileo mission was where is the water on Jupiter?'
Fran Bagenal, New Horizons mission

Above: The Pioneer 11 mission returned images like this one, which clearly pinpoint the Great Red Spot in Jupiter's south polar region.

Galileo was also the first mission to explore the extraordinary moons of Jupiter in any detail, helping us see for the first time that Jupiter lies at the centre of an extraordinary system with more natural satellites than any other planet. Holding 79 moons in its orbit (at the time of writing), 12 of these satellites have been discovered since the beginning of 2017. It is almost certain that the number will rise in the coming years, too, as we are able to explore the Jovian system in even greater detail. The vast majority of Jupiter's moons (63) are not much more than fragments of rock, none bigger than 10 kilometres in diameter, captured asteroids that now orbit the planet rather than the Sun. But at the other end of the scale Jupiter has four giant moons. Fiery Io is the most volcanically active world in the Solar System, its surface littered with lakes of lava, while icy Europa is at the other extreme, with its frozen surface protecting a global ocean of water beneath. Marbled Ganymede is Jupiter's largest satellite and the only moon we know of that has its own magnetic field, while Callisto – the outermost of Jupiter's Galilean moons – is also thought to be home to a subsurface ocean. Each moon has a unique, intriguing character, and they are all visible to us here on Earth using nothing more than a pair of binoculars. (We'll come back to these moons later in the chapter.)

Then in August 2011, NASA launched its latest mission to Jupiter – Juno. After travelling for five years, the spacecraft reached Jupiter in July 2016 and entered into a perilous elliptical orbit that took it from 8 million to just 4,000 kilometres above the cloud tops of Jupiter's surface. Juno is a mission designed to probe for Jupiter's origin and evolution, which it does by carrying instruments to measure the planet's composition and its magnetic and gravitational fields, allowing it to build up a detailed picture of the interior. It also has the best camera we've ever pointed at Jupiter, which since 2016 has been sending back the most incredible images of the giant planet ever taken. We'll let Heidi Becker, a Jet Propulsion Laboratory physicist and the radiation monitoring investigation lead on the Juno mission, explain the detail of this incredible piece of technology. 'The star tracker is an 18-pound camera, because it's passed with tungsten in order to keep the radiation out ... what JunoCam's images have shown us is that Jupiter is like going to an impressionist art gallery. It's seeing Jupiter and those atmospheric features up close, with the highest resolution that has ever been achieved. We're seeing the zones and belts, the jet streams of Jupiter, in more detail than ever before.'

For all the majesty and wonder these probes have brought us, Jupiter remains a vast distant planet. Sitting on average 588 million kilometres away from us here on Earth, it's easy to think of this world as a benign beauty, a jewel to marvel at. But what makes Jupiter the true king of the heavens is not just its beauty but its influence over the unfolding story of the Solar System. As we've explored this world ever closer we are discovering that Jupiter's effect on the fates of entire planets is far more direct and violent than we ever imagined. Its massive gravitational field exerts control not just over its moons but also over every asteroid, every world and every life force in the Solar System – whose destinies it still holds in its vast reach.

Top left: This composite image reflects the pictures seen by scientists as Pioneer 10 plunged closer to Jupiter, exposing this mysterious planet in all its glory.

Top right: During its orbit of Jupiter, Galileo recorded images of the planet's moon, Europa.

Middle right: Photographed by Galileo, Io displays its volcanic geology with a clearly visible eruption at the centre of this image.

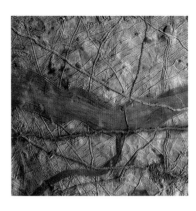

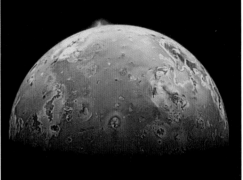

Bottom left: Galileo was the first mission to explore the moons of Jupiter in such detail.

Bottom right: Pioneer 10 completed the first mission to Jupiter in 1973, whetting the appetites of scientists to further explore this planet.

'The Great Red Spot was imaged by JunoCam. And it's our closest, most beautiful view of that incredible storm that's existed for hundreds of years that we've been watching. And the most startling thing that we saw from JunoCam's imagery was that the poles of Jupiter are actually blue. They're nothing like the orange and white planet that we grew up thinking about as children.'
Heidi Becker, Juno mission

This page: JunoCam returned images to Earth that were worthy of exhibit in art galleries, displaying Jupiter's intricate cloud patterns and turbulent weather systems.

EVIDENCE ON EARTH

To witness the awesome power of Jupiter we don't need a billion-dollar spacecraft, a rocket or even a telescope, all we need to do is to look no further than the ground beneath our feet. The surface of our planet is littered with the evidence of our distant godfather's meddling. Written across our planet in the violent gravitational handiwork of Jupiter are, at the most recent count, 190 visible impact craters, not to mention the thousands more that have long disappeared as the active geology of the Earth's surface has continually renewed itself.

One of the most famous is the Meteor crater (also known as the Barringer crater), in the northern Arizona desert, which measures more than a thousand metres across and at its deepest point drops 100 metres below the crater rim. In the late nineteenth century, geologists believed this vast feature had been created by the San Francisco volcanic field that lies just a handful of miles to the west. But in 1903, Daniel Barringer, a mining entrepreneur and sometime hunting friend of Theodore Roosevelt, began an extensive investigation of the crater and published a set of findings suggesting that it had not been created by a volcano but rather by a large incoming projectile from outer space.

This was the first time anyone had suggested such a theory, and Barringer backed it up with a host of geological evidence, including the discovery of around 30 tonnes of large iron oxide fragments from the suspect meteorite. At the time there was much scepticism that an incoming meteor could create such a massive

'Barringer's crater was formed in about 10 seconds when a meteorite struck Earth and penetrated 50 storeys of rock, exploding under the surface, uplifting layers of rock and debris, ejecting millions of tonnes of debris.'
Jeff Beal, Meteor Crater Enterprise

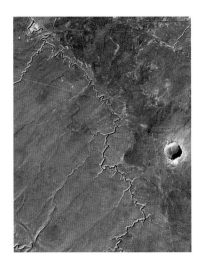

Above and opposite: Thanks to Daniel Barringer's business sense, the Meteor crater in Arizona is still owned by the Barringer family today, and is open to the public as the best-preserved meteorite crater on Earth.

Right: Impact debris from meteorites striking Earth has been found all over the planet; this sample was found in the Chicxulub meteorite in Mexico.

geological feature, but ignoring the overwhelming scientific consensus Barringer put his money and reputation firmly where his mouth was and formed the Standard Iron Company. This company was created with the sole aim of mining the crater for the 100 million tonnes of iron ore that Barringer believed lay in the ground, having been deposited by the incoming meteorite.

Even at 1903 prices, if he'd been right that amount of iron ore would have been worth over a billion dollars, so, unsurprisingly, he continued to dig down to depths of over 400 metres in his pursuit of the meteoric remains. Unfortunately for Barringer, it would take him 27 years of mining to find out what we know today. The meteorite that impacted in that Arizona desert was not only much smaller than he predicted, with current measurements suggesting it was around 300,000 tonnes (300 times less than he had thought), but also – with the benefit of advanced studies in such impacts since then – most of the nickel–iron meteorite that fell was vaporised as it struck the ground at around 43,000 km/h. Barringer's money evaporated not quite as quickly as the meteorite, but by 1929 operations at the site had stopped and most of his fortune had gone.

It would be 50 years before Barringer's work would be fully and finally verified with a landmark publication in 1963. A paper by Eugene Shoemaker and colleagues (more of him later in this chapter) analysed the similarities between the geology of the Barringer site and the craters that had recently been created as a result of testing nuclear weapons in Nevada. Shoemaker was able to show that the shocked quartz (also known as coesite) that was identified at Barringer crater could only have formed at the enormously high pressures and temperatures created by a nuclear explosion or, in the case of the Arizona crater, by the extraordinary impact dynamics of a 300,000-tonne meteorite crashing into the ground.

We now know that the giant rock that created the crater, like most of the meteorites that strike the Earth, was once an asteroid – just one of the millions of rocks that circle the Sun between the orbits of Mars and Jupiter. Most of these have been in a stable orbit for billions of years; the rocky leftovers of the birth of the Solar System, they orbit in an eternal junkyard of planetary castoffs. But occasionally one of these huge rocks is propelled towards the inner Solar System, and more often than not we think the nudge that causes it comes from Jupiter. Why this should happen and how Jupiter exerts its power not just over the asteroid belt but over all the worlds around our sun is a story we are only just beginning to understand.

DAWN OF KNOWLEDGE

From our home here on Earth it's hard to imagine the vast reaches of space that extend beyond our little bubble and the huge expanse of time over which events in the cosmos play out. But all the evidence suggests that around 4.5 billion years ago Jupiter's orbit began to change, and as it did so it triggered a period of unprecedented violence, during which the young Solar System was utterly transformed. We know this because the marauding planet left a trail of destruction in its wake that we have only just begun to explore.

As we've attempted to uncover the 5-billion-year history of our Solar System, we've repeatedly looked to the eight planets and the moons that orbit them to provide clues that fill in the gaps in our story. In the last 50 years dozens of missions have crossed our thin blue line on the way to those worlds, but only once have we sent a spacecraft to explore a very different type of world, to search amongst the rubble of our Solar System's earliest days and explore the handful of worlds that never quite made it. Sometimes the answers lie in the gaps.

Launched on 27 September 2007, the Dawn space probe took off from Cape Canaveral on a mission to explore the two largest objects in the asteroid belt – Vesta and the dwarf planet Ceres. This $450-million project was part of NASA's low-cost Discovery programme, one of a series of missions designed to be faster and cheaper to deliver. But this was no B-list mission for NASA. Equipped with the most advanced ion-propulsion system ever used in outer space, Dawn would be able to reach the asteroid belt and become the first craft in the history of human exploration to enter orbit around Ceres. And yet without a major planet in its itinerary the attention of the world only seemed to glance towards this mission, and for the most part its extraordinary findings slipped between the gaps of public attention.

The aim of the Dawn mission was simple: this would be the first spacecraft to travel to the asteroid belt, the first to orbit an object in this uncharted region of the Solar System, and the first to visit a dwarf planet. All in all, this mission profile was designed to allow scientists a rare chance to travel deep into the past and glimpse the earliest moments in the origin of our Solar System. The mission allowed us to explore the evolution of Vesta and Ceres – two very different objects, frozen in time for billions of years since they faltered and failed in the process of becoming fully-fledged planets.

To travel to the asteroid belt would be a slow but steady voyage for Dawn. After hitching a conventional ride into space on the Delta II rocket, the spaceship began firing its thrusters to orientate and propel itself away from the Earth and towards Mars. The ion thrusters on Dawn were an efficient but gentle form of propulsion for the spacecraft, taking four days at full throttle to accelerate the craft from 96km/h. It's not exactly a sports car, but when it comes to efficiency the ion propulsion system is difficult to beat. Powered by a 10kW solar array, Dawn's engines ran for almost six years across the 11-year lifetime of the mission, propelling the craft across space using less than 400 kilograms of its xenon fuel supply, while reaching speeds of 41,360km/h. Its trajectory took it on a 16-month journey to Mars, where a gravitational kick sent it on its way out to the asteroid belt and onwards towards Vesta, its first destination.

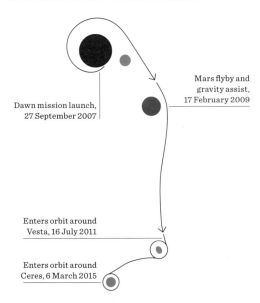

DAWN'S TRAJECTORY TO UNCHARTED SPACE

Dawn mission launch,
27 September 2007

Mars flyby and
gravity assist,
17 February 2009

Enters orbit around
Vesta, 16 July 2011

Enters orbit around
Ceres, 6 March 2015

Opposite: Dawn was launched in September 2007, on a mission to travel to the asteroid belt and orbit an uncharted region of the Solar System.

'8:36 am: Dawn is on its way to the asteroid belt.'
NASA Dawn launch

BOUNDARY BETWEEN WORLDS

'The asteroid belt holds a remarkable amount of information about the Solar System's dynamic and dramatic evolution.'
Konstantin Batygin, astrophysicist

Almost 500 million kilometres from the Sun, the asteroid belt lies halfway between Mars and Jupiter, the boundary line between the four rocky worlds that orbit closest to the Sun and the giant gas and ice planets that lie beyond. Made up of millions of pieces of rock from the dawn of the Solar System, it's a graveyard of failed planets, remnants from the heyday of planetary construction.

We often imagine the asteroid belt to be a dense, unnavigable mass of tumbling rock, but in reality it's something quite different. The total mass of all those rocks is estimated to be 3×10^{21} kilograms, which although an enormous-sounding number is just 4 per cent of the mass of our moon, with a third of that mass found just in the largest object, Ceres. The rest of this cosmic junkyard is made up of objects covering a vast range of different sizes; on one end of the scale we think there are at least a couple of hundred asteroids larger than 100 kilometres in size, then come a million or so stretching to at least 1 kilometre or more, and a countless spread of smaller objects number in the millions and billions depending on how small you want to go.

All of this sounds like it should make for a crowded environment full of collisions, the perfect location to form the basis of a computer game or disaster movie. But even with that many objects, the space they occupy is so unimaginably vast, spread out across 13 trillion, trillion cubic miles of space, making the average

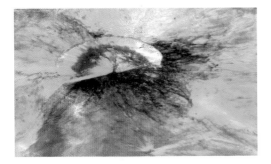

distance between any two large asteroids approximately 2 million miles, or eight times the distance from the Earth to the Moon. These distances are so large that when we send spacecraft to the outer Solar System the chances of hitting anything are so small that the engineers don't need to make any course corrections to pass through. In fact, if you were standing on an asteroid you'd be lucky to see another one – and if you did it would appear as little more than a tiny point of light in the darkness. As it travelled through this vast emptiness in the early summer of 2011, the Dawn spacecraft slowly began to bring its first target into view. From a distance of over 1.2 million kilometres, Dawn took its first photograph of Vesta against the backdrop of a thousand stars. Over the next two months Dawn homed in on Vesta, chasing it across the vast reaches of space to arrive at this, the second-largest rock in the belt. Entering its orbit on 16 July 2011, Dawn became the first spacecraft to enter within the gravitational hold of an object in the asteroid belt, and for the next 14 months it would explore this extraordinary ancient object, providing a glimpse of our Solar System in the very earliest days of its life.

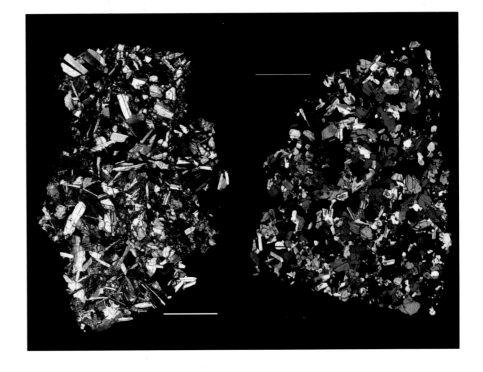

Opposite: Artist's concept of NASA's Dawn spacecraft, pictured with Vesta (left) and Ceres (right).

Above top and middle: Craters on Vesta; Antonia (top) lies in the enormous Rheasilvia Basin in the southern hemisphere, while Sextilia sits 30 degrees south latitude.

Above bottom: The giant asteroid Vesta and its heavily cratered surface, shown here photographed by Dawn and coloured to show presence of minerals.

Bottom right: Thin slices of HED (howardite, eucrite and diogenite) meteorites – recovered in Antarctica, which are believed to have originated from Vesta.

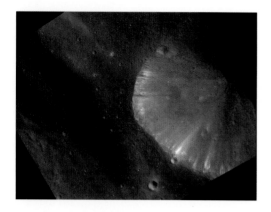

GEOLOGICAL TIME SCALE OF VESTA

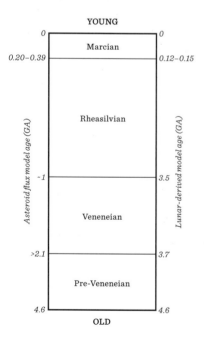

Above: The geologic time scale of asteroid Vesta as derived from geologic mapping. Time is measured in Ga, billions of years before the present.

Bottom: The southern hemisphere of Vesta, showing the Rheasilvia crater.

When 4 Vesta (to give it its full name) was first discovered by the German astronomer Heinrich Olbers in 1807 it was the fourth asteroid to be discovered after Ceres, Pallas and Juno (hence the 4 in its formal designation). Olbers was convinced he had found the remains of a lost planet, destroyed in some long-ago collision. Even so, just like the other three objects, at first Vesta was given full planetary status, bringing the number of classified planets in the Solar System at that time to 11. It wasn't until 1845, nearly 40 years later, that a flood of further discoveries began that quickly increased the number of objects in this region to a point where it was clear they couldn't be classified as planets, and so in the 1850s, 15 planets became eight and the classification of these objects as a different class of minor planet, asteroids, had begun.

One hundred and sixty years later, as Dawn began its year-long residency in the vicinity of Vesta, it soon became apparent that Olbers's suggestion that this was a fragment of a long-lost planet was not so far off. Vesta is a spheroid-shaped rock, 530 kilometres in diameter and pockmarked with impact craters. The most striking example of these is the crater Rheasilvia, which at 500 kilometres wide and 20 kilometres deep is one of the largest impact craters in the Solar System. At its centre is a mountain that rises into the blackness, 22 kilometres from the crater floor, making it the second-tallest peak in the Solar System after Mars's Olympus Mons. What's particularly extraordinary about this crater is that we have long suspected that a particular class of meteorite found on Earth, known as HED meteorites, actually come from material that was blown off the surface of Vesta when this impact occurred around a billion years or so ago. Using Dawn's instrumentation we were able to directly measure the geology of rocks in Rheasilvia, and when compared with the samples on Earth we were able to conclude that they are of the same origin. In one of the most beautiful examples of geological analysis in the history of the subject, we now know that we can hold a piece of this far-away asteroid in our hands and directly analyse the secrets hidden within it.

When taken all together, the HED meteorites and the findings from the Dawn spacecraft have enabled us to paint a detailed picture not just of Vesta today but also of its history. Dawn has revealed to us that Vesta has a metal-rich core, just like all of the terrestrial planets – including Mercury, Mars, Venus and Earth – and analysis of its surface geology has enabled us to date the formation of the asteroid to a couple of million years after the first solid objects began to take shape in the Solar System. These are just two of the clues that point us to an extraordinary conclusion – Vesta is perhaps the last remaining example of a protoplanet, a half-formed world from the very earliest days of the Solar System. Frozen in time, this is a stage that every world we know of passed through, including the Earth. Vesta was on the way to becoming a planet almost 4.5 billion years ago, but something happened to stop its progress; while other worlds went on to flourish and other protoplanets disintegrated in the violent collisions of the early Solar System, Vesta somehow seemed to survive, a fossil of planetary formation for us to not only look at but, through a quirk of fate, actually hold. But this was not the end of Dawn's dive into our deep history; departing Vesta after 14 months of study, Dawn fired its ion drive, leaving the orbit and heading off for its second planetoid rendezvous at the largest of all the asteroids, Ceres.

Top: Map of Vesta showing the distribution of dark materials throughout the asteroid's southern hemisphere. The circles, diamonds and stars pinpoint dark material in craters, spots and topographic highs. The dashed line depicts the rim of the Veneneia basin, the black line the rim of the younger Rheasilvia basin. The red and white indicate high topography and blue and violet indicate low topography.

Bottom: Geological map of Vesta. Brown denotes the oldest, most heavily cratered surface; purple in the north and light blue the terrains modified by the Veneneia and Rheasilvia impacts. Light purple and dark blue below the equator show the interior of the impact basins. Greens and yellows represent recent landslides or other downhill movement and crater impact materials, respectively.

JUPITER'S RED SPOT

The outer planets of our Solar System are a paradise for those of us studying the atmospheres of different worlds, because they represent worlds without complicated topography. There are no mountain ranges, no valleys, no continental boundaries to get in the way of the perfect fluid dynamical flows. So in a way looking at the weather systems on the giant planets, those perfect jet streams, the large spinning vortices and the evolving cloud patterns is what our world might look like if you didn't have the complicated continents, valleys, mountain ranges, getting in the way of those weather systems. These systems can persist and continue potentially for very long periods of time. The Great Red Spot on Jupiter is a wonderful example of a storm system that's been raging for more than a century. How long they last, we don't know, because our exploration of the Solar System has been limited to just the last few decades. This makes it an open question as to what is going to cause the dissipation and the shifting of these weather patterns that we see on the giant planets.'

Leigh Fletcher, Jupiter Icy Moons Explorer mission

Opposite and bottom:
Jupiter's Great Red Spot is a
gigantic storm system, about
twice the size of Earth, which
has raged for over a century.

Top: The Great Red Spot
storm system is visible on
these two images of Jupiter
as the bright white area.

CERES'S SECRETS

'When Dawn arrived at Ceres, one of the most exciting things was when we started to see the surface. What had been a blobby disc in telescope data became a real landscape, with features, available for us to explore.'
Bethany Ehlmann, planetary geologist

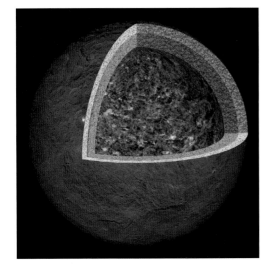

Dawn arrived at the dwarf planet Ceres in early March 2015, and from the very first moments of its observations it sent back some astonishing images. Very little was known about Ceres before our spaceship arrived, but it soon became clear that this was a rock with far more of a story to tell than we could ever have imagined, a story that would help reveal a new perspective on the deep history of our Solar System and the enormous influence Jupiter has exerted over it.

The Italian astronomer Giuseppe Piazzi discovered Ceres on 1 January 1801, and at first he reported his findings as the discovery of a new comet in the region of space between Mars and Jupiter, but it seems from the start that all of his instincts told him this was something else. In announcing his observations in a letter to two fellow astronomers, he noted that 'since its movement is so slow and rather uniform, it has occurred to me several times that it might be something better than a comet'. Piazzi and his colleagues speculated that rather than a comet they might have found something even more spectacular – a new planet. But after just 41 days of observation, Piazzi fell ill, the path of the newly discovered object strayed into the direct glare of the Sun, and in a twist of high astronomical drama the newly-found planet Ceres was lost.

With just a tiny number of observations across a small piece of sky, calculating Ceres's orbit was deemed impossible using the mathematics of the time, and many contemporary great mathematical minds predicted it might never be found again. It would take the work of a young German mathematician, Friedrich Gauss, to solve the problem. Attracted to the 'difficulty and elegance' of using mathematics to compute the orbit from just a few coordinates, Gauss pioneered a novel mathematical method that allowed him to calculate the path of Ceres using just three of Piazzi's original observations. On the very last day of 1801, two astronomers using Gauss's methods searched the region of the sky to which the calculations guided them, and to everyone's delight and relief recovered Ceres from the darkness. Gauss became a mathematical superstar and Ceres, firmly back in our sights, was stamped with the classification of a planet. For the next 50 years, Ceres and its companions – Vesta, Pallas and Juno – remained listed as planets until, as we have seen, the discovery of a raft of other objects in this region of space gave rise to our identification of the asteroid belt, and in 1852, Ceres was demoted from planet to asteroid number one.

For the next 160 years little was known about this asteroid orbiting the Sun at a distance of 413 million kilometres. We knew that it was the biggest object in the asteroid belt and the only object amongst the millions of asteroids that was large enough to have been rounded by its own gravity. We could also make a good approximation of its size (at around 1,000 kilometres in diameter) and composition (rock and ice, what else?!), but when it came to surface features even the powerful eye of the Hubble Space Telescope could make out only the vaguest of features, intriguingly strewn across its surface. It would take Dawn's arrival for us to truly understand the complexity of this world and its history.

Even in the early approach phase of the mission, Dawn was able to capture a quality of imagery that at once gave a new perspective on the long-hidden surface. It was no surprise to find it heavily cratered, but while speeding towards Ceres, Dawn beamed back a series of mysterious images, revealing a number of bright spots on the surface of what we had thought would be a dull, dead and frozen world – not what you

Opposite: Salar de Uyuni, in Bolivia, is the largest salt flat on Earth at 10,582 square kilometres.

Above: Artist's impression of the layers of Ceres; data from the Dawn mission suggest an internal clay-like layer, encased in a thick ice, salt and mineral layer, which sandwiches a briny, salty layer.

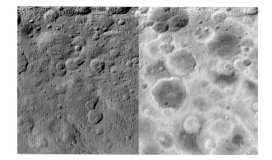

would expect on a planet that was assumed to be entirely geologically inactive. The brightest of all of these highly reflective areas was imaged near a massive crater known as Occator. At 92 kilometres wide and 2 kilometres deep, it is one of the biggest on the surface of Ceres. Spot 5, as it became known, is in the centre of the crater and became the focus of intense investigation as Dawn began its three-and-a-half-year residency in orbit around Ceres. The crater is thought to have formed from a large-scale impact around 80 million years ago, and so the presence of what appeared to be geological active material on the surface was particularly surprising and at first left the mission scientists perplexed. Not only that, but a haze was also seen to periodically appear above Spot 5, suggesting something was going on beneath the surface.

After detailed analysis using Dawn's full range of imaging and spectroscopic capabilities, we now believe that Spot 5 is the result of a geological hotspot – the brightness created by the highest concentration of carbonate minerals ever found outside the Earth in a dome structure right in the centre of the crater. Why does this matter? Well, we know that this type of salt can only be formed in the presence of liquid water, and it's almost certain that an impacting asteroid could not have delivered such material to the surface of Ceres, so the upwelling in which it is found suggests that it comes from deep within this world. It now seems that Spot 5 is

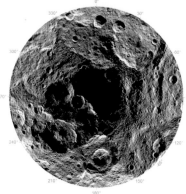

Top: Kwanzaa Tholus on Ceres; a tholus is a small mountain, and this is 35 by 19 km wide, and 2 or 3 km high.

Bottom: Using data derived from Dawn's mission, scientists have been able to create images of the north and south poles of Ceres.

Below: Salt crystals like these on Earth may contribute to Ceres's shining spots.

Planetary geologist Bethany Ehlmann on Occator

As Ceres came into view and the spacecraft was coming in, we saw, rotating into field of view, these amazing bright spots on the surface, pot marking it, here and there, nothing like we'd ever seen before. And the most outstanding ones were found in Occator crater.

Occator is about a hundred-kilometre crater or so, and it's one of the larger, but younger, craters on Ceres, formed within the last hundreds of millions of years, maybe a billion years. The most intriguing thing, though, is what we see in the centre, which is a central pit and then a bright dome of this super-bright material.

The bright spots are the signs that Ceres, pretty recently, had liquid water or, at the very least, the products of liquid water have managed to extrude to the surface. This is incredibly exciting, because it shows Ceres is an active world, where cryovolcanism is bringing these briny materials to the surface.

the remains of an ice volcano, one that is perhaps still partially active, and thus able to create the haze we have observed. The two brightly coloured patches inside it are thought to be the salty residue left by this ancient formation, deposits of sodium carbonate from an erupting ice volcano that may have first been triggered by the impact that created the crater. However, the dome-like shape upon which the deposits sit is almost certainly evidence of geological activity below the surface, suggesting that hydrothermal activity was involved in bringing the salts to the surface.

And Spot 5 is far from isolated, we've now seen areas like the ones in Occator all over the surface of Ceres, and this begins to paint a tantalising picture of Ceres's past – a past where heat from within the planet created a subsurface ocean, or at least localised pockets of water that surged upwards in the form of great ice volcanoes, bringing with them on the asteroid the carbonate salts that we see sparkling today.

As Dawn delved deeper beneath the surface the surprises kept coming thick and fast. Dawn was able to accurately measure the overall density of Ceres and found it to be 2.16 g per cubic centimetre, which is less than you'd expect for a rocky body. Ceres is about two-thirds the density of our Moon, which suggests that as well as rock, the asteroid must also hold significant amounts of water ice.

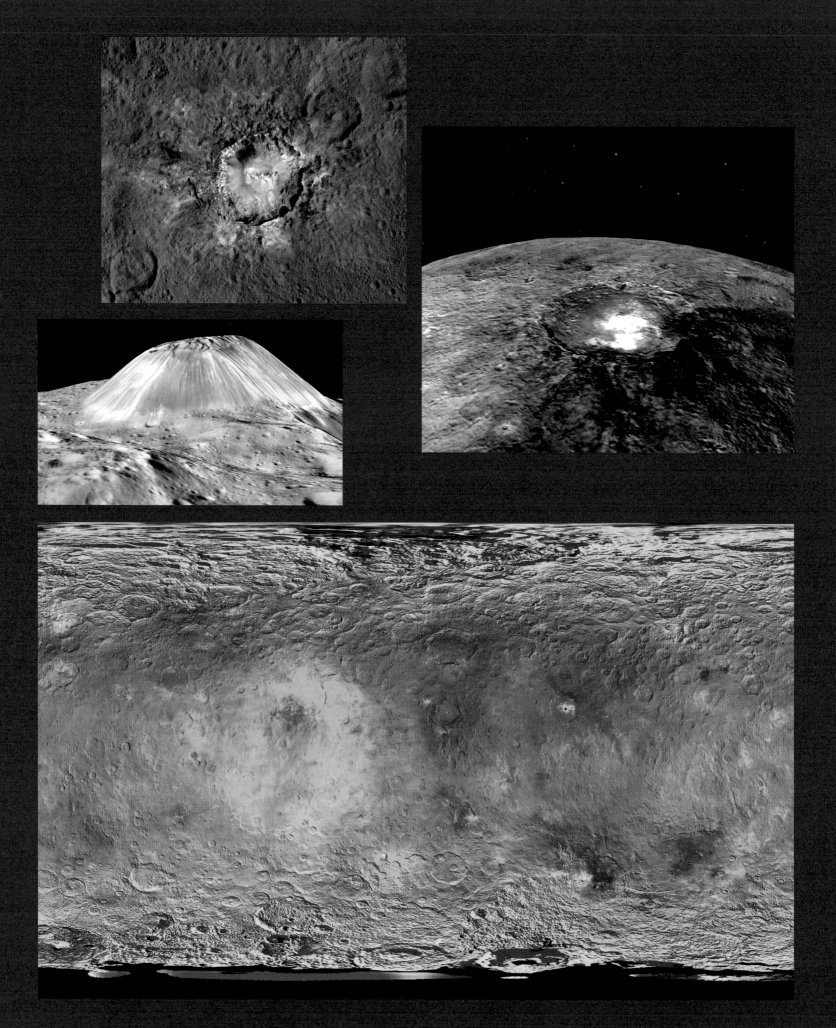

We now suspect that at this early time in the life of the Solar System, as Jupiter circled the young Sun, something caused the giant planet to spiral inwards towards the Sun, ploughing right through the asteroid belt.

This had been suspected long before Dawn arrived, but with our eyes and ears now at work, as Dawn circled Ceres it was able to take precise measurements and so uncover anomalies in its orbit that revealed something remarkable about this lump of rock and ice. By carefully mapping the shape and gravitational field we found that the rock and ice weren't evenly distibuted. It appears that Ceres's interior is nothing like that of other asteroids in the belt; instead it appears to have a rocky core surrounded by an ice-rich crust. This is not the jumbled structure of an asteroid; the 'differentiation' into layers that was discovered by Dawn is something we only see in planets. It seems that around 4.5 billion years ago, Ceres had taken its first steps to becoming a fully-fledged world, and then something went catastrophically wrong.

We now think that 4.57 billion years ago a young, rocky planet began to form in the region we call the asteroid belt. In its infancy this protoplanet was a water world, warmed from within by the heat left over from the violence of its birth. Even in this frozen region of space, orbiting 413 million kilometres away from the weak young Sun, a deep saltwater ocean encircled the young Ceres, protected from the freezing temperatures of space beneath a thin layer of ice. Across this ancient embryonic world ice volcanoes broke through, transforming and renewing the surface as it was bombarded by the bodies that swarmed around it.

Today we see just the lifeless remains of this geologically active world in the distinct mountainous landscape and sparkling salts that clothe it, but the subsurface ocean is long gone, locked away as a thick layer of ice between the surface and the core. This is a nascent planet that has been in deep stasis for billions of years; frozen in time, it is a precious piece of history, an artefact that allows us to glimpse our origins and understand just how precarious a time this was for every planet, including our own. We now think that before Ceres could fulfil her potential, a great disturbance sent shockwaves through what would eventually become the asteroid belt and cut its young life short. The vast amounts of rock and ice that made this part of the young Solar System such a rich breeding ground for the birth of planets was wiped clean of almost all of its building material, leaving just a fraction of the material needed to build a planet and nothing from which Ceres could grow. Computer simulations suggest that originally the asteroid belt contained enough material to build a planet the size of Earth, and perhaps that was once Ceres's destiny perhaps it was even destined to become a subterranean water world capable of supporting and nurturing life. But before any carbon chemistry could spark into seeds of life, Ceres's fate was sealed, triggered by the wanderings of a giant that lay beyond in the darkness.

We now suspect that at this early time in the life of the Solar System, as Jupiter circled the young Sun, something caused the giant planet to change course and spiral inwards towards the Sun, ploughing right through the asteroid belt. The resulting gravitational chaos scattered the contents of the asteroid belt far and wide. Embryonic planets were knocked out from their orbits, the building material of these planetesimals was thrown out and in towards the Sun. Jupiter's charge displaced 99.9 per cent of the material in this region of space, leaving Ceres half-formed, half-built, but with nothing more from which to grow. In this part of the Solar System at least, the planetary construction business was shut down for good.

SUPER EARTHS

Our Solar System is not typical at all;
it may in fact be very odd indeed.

ow do we know such a dramatic turn in the Solar System's history took place? The greatest evidence has come not from our own system but from the study of other star systems and their planets, a field of science that has exploded in the last 30 years. Until the twentieth century the only evidence of how planetary systems formed and functioned came from one single source – Earth's Solar System. Our own backyard was the one place that we had the technology to study, and that meant our view of how a planetary structure worked was deeply skewed towards the seemingly simple structure of a star at the centre, four small rocky planets orbiting it, then a ring of millions of rocky pieces, remnants from its formation, followed by a series of four more planets, the gas giants. It has a pleasing regular structure about it – all the rocky planets sitting close to the star and the gas giants orbiting further out. We supposed this must be the natural order of things, the way that solar systems have to form, but that all changed when the technology arrived to allow us to detect planets around distant stars. It soon became clear that our Solar System is not typical at all; it may in fact be very odd indeed.

In most of the planetary systems we've observed, the set-up is remarkably different to our own. As we have extended our gaze and explored the structure around other stars with the help of exoplanet-hunting technology like the extraordinary Kepler Telescope, we have begun to see that one of the most commonplace categories of planet is completely missing from our own. Inward of Mercury, our Solar System is essentially an empty void, but look at the equivalent region around most stars we have studied and we often find it packed with a completely different class of planet – Super Earths. These worlds are between two and eight times the mass of the Earth, with thick hydrogen-rich atmospheres, they sit incredibly close in to their stars, orbiting at a rate that makes the head spin.

The first discovery of a Super Earth around a main sequence star like our own was made back in 2005 by a team led by Eugenio Rivera, using data from the Keck Observatory in Hawaii. Gliese 876d is an exoplanet almost seven times the size of the Earth which is orbiting just 3 million kilometres away from its red dwarf star, Gliese 876. Thought to be a rocky planet with surface temperatures estimated to approach 341 degrees Celsius, with an orbital period (year) of just under two

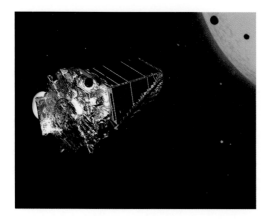

Above: An artist's impression of NASA's Kepler Space Telescope on its K2 mission, which enabled the discovery of over 100 new planets.

Right: The NASA Infrared Telescope Facility at the Mauna Kea Observatories, in Hawaii, where the first Super Earth was discovered.

ORBITS OF THE PLANETS OF GLIESE 876

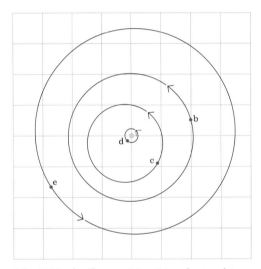

2 June 1997: each grid square = 0.1 au x 0.1 au; planets and star
not drawn to scale.

days, this is a world completely inhospitable to life. To understand why this rocky planet is found so close in to the Sun we need to understand the rest of the planetary system, and in the case of Gliese 876 that means looking at the three other planets circling further out from the star.

The first of these planets to be discovered, and one of the earliest-ever observations of an exoplanet, is 876b. Discovered in 1998 by the legendary planet hunter Geoff Marcy, we believe this world is a gas giant at least twice the size of Jupiter but orbiting much closer to its parent star. In fact, at just 0.2 au (30 million kilometres) this massive planet is orbiting much closer in than Mercury orbits the Sun. At that distance 876b zooms around on its orbit once every 61 days, compared with Jupiter's stately 12-year journey around the Sun. Either side of this giant planet are two other worlds: 876c, a gas giant three-quarters the size of Jupiter; and 876e, a planet we think is about the same size as Uranus. All four of these worlds orbit within a distance less than the orbit of Mercury – it's as if the whole planetary system has been squashed, packed in tightly around the star, and this configuration is far from rare. As we have explored more and more planetary systems we've discovered this combination of Super Earths and close-orbiting gas giants is the nearest to a normal layout we observe.

What underlies this is the fact that planetary systems are not the static and stable systems we had always assumed them to be. Far from the heavens being full of planets following eternal clockwork rhythms, most systems – including our own – are chaotic, dynamic and turbulent. The reason for this is that particularly early on in the infancy of a new system, planets frequently migrate, forming in one part of the system before wandering off into different zones. There are a number of different mechanisms behind these wanderings, but perhaps the most formative on the evolution of a planetary system is the Type II migration that occurs to gas giants early on in their lives.

A gas giant planet like Gliese 876c would have almost certainly begun its life far further out than it is today, in a region more akin to where we see Jupiter or Saturn in our own solar system. At the very earliest stages of its existence, as this newly formed planet orbited within the cloud of gas and dust of its formation, the planet began to clear a path in its wake, and in doing so triggered a process that would transform its orbit. A shift in its gravitational interactions with the gas-rich disc would have slightly reduced its angular momentum and sent the planet on a different course, spiralling inwards towards the star at the centre of the system. This seems to be a common path trodden by many gas giants around distant stars, and is the reason we find a preponderance of so-called 'hot Jupiters' – massive gas giants sitting so close to a star that they orbit it in less than ten days. The effects of these migrations has an impact far greater than just this, though; in ploughing inwards a migrating gas giant like Gliese 876c ends up clearing vast areas of space, halting any planetary formation and instead pushing material inwards towards the star, creating the perfect scenario for Super Earths (like 876d) to form.

This cosmic pinball is the reason why so many of the planetary systems we have been able to explore follow the pattern of hot (ish) Jupiter and Super Earths, but why not ours? What was different about the history of our Solar System that led to the more expansive layout we see today? Scientists suspect the answer lies with our own gas giant, Jupiter, and the journey it took across the young Solar System.

Top: The Gliese system as seen from Earth by the Oschin Schmidt Telescope, near San Diego, California.

Bottom: The orbits of the planets of Gliese 876. Note that the strong gravitational interactions between the planets causes rapid orbital precession, so this diagram is only valid at the stated epoch.

THE WORLD ACCORDING TO JUPITER

'Jupiter's motion through the early Solar System can be thought of as quite destructive, but in destruction there are also opportunities for creation.'
Leigh Fletcher, Jupiter Icy Moons Explorer mission

The story of Jupiter begins not in the far reaches of the Solar System, 778 million kilometres from the Sun where it orbits today, but much closer in. We now think that almost 5 billion years ago Jupiter formed around 520 million kilometres (3.5 au) from the Sun.

At this time the young giant was still surrounded by the disc of gas and dust from which it had arisen, but Jupiter began to slowly but surely clear its path through the cloud and debris. Over a relatively short period of time of just a few million years, the path through its orbit became completely clear and, just like the huge gas giants we have observed around alien stars like Gliese 876, this caused Jupiter to start spiralling inwards towards the Sun. This process was driven by the gravitational interaction between Jupiter and the cloud of gas, which as it fell into the Sun itself effectively dragged the planet with it. This was no subtle shift in Jupiter's orbit. While it spiralled inwards it would keep falling closer and closer to the Sun, passing through the region of the asteroid belt and robbing planetary hopefuls like Ceres and Vesta of the material they needed to complete the journey to planethood.

But this was just the start of Jupiter's reign of terror; further and further the young Jupiter fell, until it reached the vicinity of where Mars orbits today, just 225 million kilometres or so from the Sun. This was no quiet visitor to our local neighbourhood, a planet of this size can wreak havoc just by its proximity, and Jupiter's arrival in the inner Solar System did just that, transforming the destiny of many of the emerging new worlds, including our own. Leigh Fletcher, from Jupiter Icy Moons Explorer mission, explains this action as a tug of war; 'between the orbits of Mars and Jupiter we have a large accumulation of objects, rocky objects mostly, known as the asteroid belt. They're not able to form and agglomerate into a complete planet because of the constant tug of war between the gravitational forces of the Sun and of Jupiter. So what we see today is a large disc of objects that could be the remnants left over from the formation of our solar system.'

To understand just how powerful a presence Jupiter must have been as it marauded around the inner Solar System, we need to look no further than the worlds it has in its iron grip today. Jupiter has (at the time of writing) 79 known moons held in its orbit by the powerful gravitational field that is generated by the planet's enormous mass, a powerful forcefield that reaches out from the planet deep into the Solar System. It's what has made Jupiter so destructive in the past, and through our exploration of the Jovian system we have also witnessed its awesome power today.

Nowhere is this power more dramatically revealed than on Io, the closest of all the moons to Jupiter. Io orbits just 350,000 kilometres from the cloud tops of its mother planet, a distance that makes it a gravitationally tortured world. A volcanic vision of hell, this is the most geologically active object in the Solar System, with over 400 active volcanoes, from which lava flows stretch hundreds of kilometres across the moon's surface, and vast clouds of pyroclastic material blow hundreds of kilometres up into space, creating distinctive umbrella-shaped plumes above its surface. The surface of Io is perhaps the closest that our Solar System has to offer of a vision of hell, and all of this activity is created not by ancient internal heat, as we see on our planet Earth, but by the power and the influence of Jupiter's gravitational pull.

Opposite: Io, as imaged by the Galileo spacecraft, on its closest flyby on 3 July 1999.

'Io is the most volcanic moon in the Solar System; every single time a mission has been there, we've seen volcanos erupting, and sulphurous material being sent up into Io's sky.'
Leigh Fletcher, Jupiter Icy Moons Explorer mission

Here on Earth the volcanism that we see across our planet is driven by a combination of the heat that remains from our planet's violent formation and that which is released from the decay of radioactive elements locked away in the planet's interior. This heat, combined with the high pressures below the surface, creates the molten rock that pools within the mantle layer of the Earth and then in certain places this magma finds a route to the surface, creating a volcano. On Io the volcanism is nothing like this; instead it's being generated by a completely different mechanism, a process called tidal heating that is driven in large part by its proximity to Jupiter and the gravitational tug of war between other nearby moons. Io is caught between the vast mass of Jupiter on one side and the three other Galilean moons – Europa, Ganymede and Callisto – on the other, a dynamic that pulls in different directions to violent effect.

It's a dynamic that we see in part on our planet every day as the force of gravity flexes its muscles in the interaction between the Earth and the Moon, creating the twice-daily ebb and flow of our tides. On Earth the gravitational tug of the Moon creates a tidal force that moves vast volumes of water around the planet, with the biggest difference between low and high tide creating a shift of a maximum of 12 metres in the height of water. The same principle also applies on the surface of the Moon, where the gravitational pull of the Earth creates tides not of water but of rock. Moon tides are small changes in the height of rocks on the lunar surface caused by the pull of the Earth's gravity. In the case of the Earth/Moon system these rock tides are measured in just a few centimetres, but Io tells a very different story.

Io is roughly the same distance from Jupiter as our Moon is from Earth, and both moons are of equal size, but because Jupiter is so much more massive, the tidal forces it generates are much greater, with the gravitational pull of Jupiter moving the bulge of rock tides up to 100 metres between low and high tide. That's five times the height of any tide here on Earth lifting the levels of solid rock up and down. (In fact, Io contains the least amount of water of any known object in the Solar System.) But this is not the end of Io's torture. Its relationship with the other large moons – particularly the next moon out, Europa – also adds to the destructive forces that are pulling the very substance of Io apart.

Io whips around Jupiter much quicker than our moon goes around the Earth (once again, due to the mass of Jupiter), completing an orbit once every 42 hours. If left alone Io would complete this orbit in a near-perfect circle around the giant planet, but Io is far from alone, and with Europa circling Jupiter precisely once every two orbits of Io it means that in every second orbit Europa and Io line up, and that rhythmic orbit has consequences. This is what's known as an orbital resonance, and the result is that it gives Io an additional gravitational kick that moves the moon's orbit out of a circle and into an ellipse. Ganymede is also in a resonant orbit with Io in a 1:4 orbital period ratio, and the outcome of this is that as Io orbits the gravitational forces are constantly shifting and changing, and that stretches and squashes the moon, creating vast amounts of friction that causes its interior to heat up. It's a process called tidal heating and it's this generation of heat that produces Io's volcanoes, raising the temperature in its interior to over 1,200 degrees Celsius and creating the violent world of lava lakes and volcanic plumes that we have witnessed through the eyes of Galileo and Voyager. The pull of the two moons on Io also stops it gradually falling inwards to Jupiter and being completely ripped apart.

Io is an extraordinary example of the power of Jupiter and a demonstration of how this planet can shape and change the destiny of worlds. Because of its huge mass and the intense gravitational force this produces, any planet or moon that gets too close to it is under threat. While Earth is at a relatively safe distance today, 4.5 billion years ago, as Jupiter moved into the inner Solar System, it began to exert its power right in the spot where Mars and our own world were beginning to form.

Left: On Io, we have discovered that the volcanic eruptions are caused by very different mechanisms than on Earth, being the result of tidal heating instead of a combination of heat and high pressure.

Top: Jupiter's moon, Ganymede, assembled as an image from data received by Voyagers 1 and 2 and Galileo.

A volcanic vision of hell, Io
is the most geologically active
object in the Solar System.

of Io, captured by Galileo.
Sodium atoms orbiting Io
scatter sunlight to create
waves. The white light

top image and on its right
edge in the bottom image
show thermal emission
from Pele.

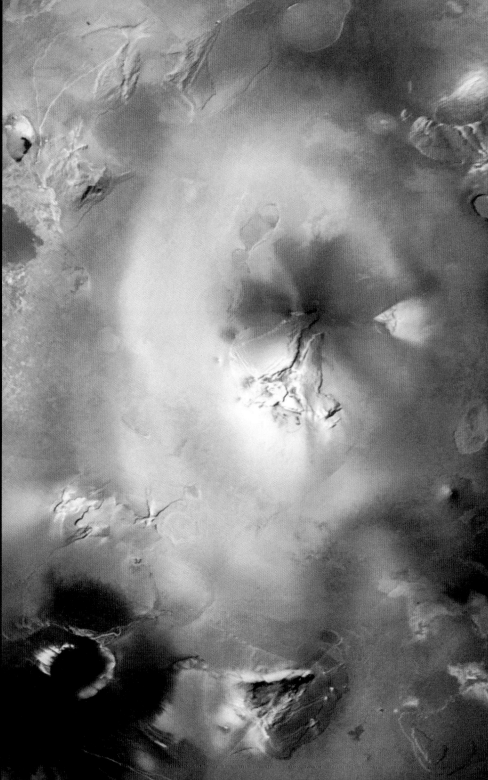

Opposite and above: Io's
volcanoes in action. The

JUPITER: CHAOS AND CREATION

Four and a half billion years ago, the region of the Solar System we call home was densely packed with gas and rock, a planetary nursery full of embryonic worlds, each with a different destiny in store. But this fertile space was about to be thrown into chaos by the Solar System's most unwelcome of visitors. As Jupiter was pulled from its distant orbit and continued its inwards descent it paused in this planetary breeding ground. The consequences of this action would be significant and far reaching.

Detailed simulations of the impact this wandering Jupiter would have made suggest that the fate of any Super Earths caught in mid-formation would have been thrown into chaos. The models suggest that Jupiter's inward journey would have driven thousands upon thousands of planetesimals, many over 100 kilometres wide, spiralling into the inner Solar System. This planetary breeding ground was turned into a battleground as Jupiter's presence threw the orbits of any newly forming Super Earths into the overlapping paths of thousands of massive blocks of rock and ice. The result was carnage, a cascade of collisions that blew apart planets and left a trail of fragmentary worlds in its wake; a chain reaction that would culminate in the formation of a spiralling swirl of gas, dust and rock that would have swept any surviving Super Earths towards the Sun and ultimately to their destruction.

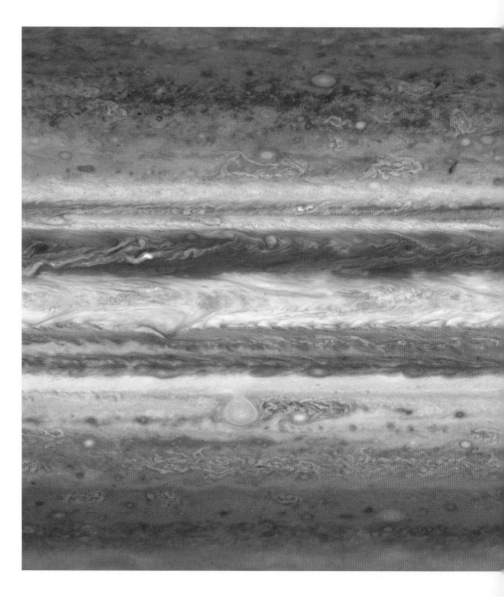

Above: The extraordinary power of Jupiter, as shown by a huge dynamic storm dominating the southern edge of the planet's north pole.

Right: The Cassini mission's flyby of Jupiter enabled scientists to produce the most detailed global maps of the planet.

Leigh Fletcher on Jupiter's influence on
planetary formation
*'Jupiter's mass is so enormous that if you have Jupiter
moving around within the early Solar System it can
really mess things up; think about what would happen
if Jupiter were to migrate inwards towards the forming
rocky planets like Mercury, Venus, Earth and Mars.
Now the presence of such a large heavy object would
mean all those forming planets would be scattered
all over the place, potentially undergoing cataclysmic
collisions that would break them apart and reform them
again. It certainly would not be conducive for the rise
of life and definitely not for the rise of complicated life
forms like our own.'.*

In what was just a fleeting moment in the history of our Solar System, its destiny changed for ever. Our Solar System would not fit the norm of Super Earths and the tightly packed gas giants that we see across the Milky Way, instead we think the inner Solar System would have been swept clear of up to 90 per cent of the rocky material thrown in by Jupiter. With just crumbs remaining, four small terrestrial planets began to form from the rubble. With Jupiter marauding in its orbit, much of the material that could have gone on to form Mars was lost, leaving the planet with enough scraps to make a planet half its potential size and transforming it from a possible habitat for life to a world that would freeze and die in its adolescence. Just that little bit further in, Earth found itself forming in a slightly more substantial ring of gas and dust that would make it the largest of all the terrestrial planets. A lucky escape for our future home, because had Jupiter continued on its inward path Earth might never have formed at all; but just as it looked like it would sweep everything away, Jupiter stopped in its tracks.

On its inward journey Jupiter had transformed the inner Solar System, sweeping away the formative giant planets and leaving just enough material for Mars, Earth, Venus and Mercury to form. But the stage was now set for another player to enter the story and shift the destiny of both our world and the Solar System once again.

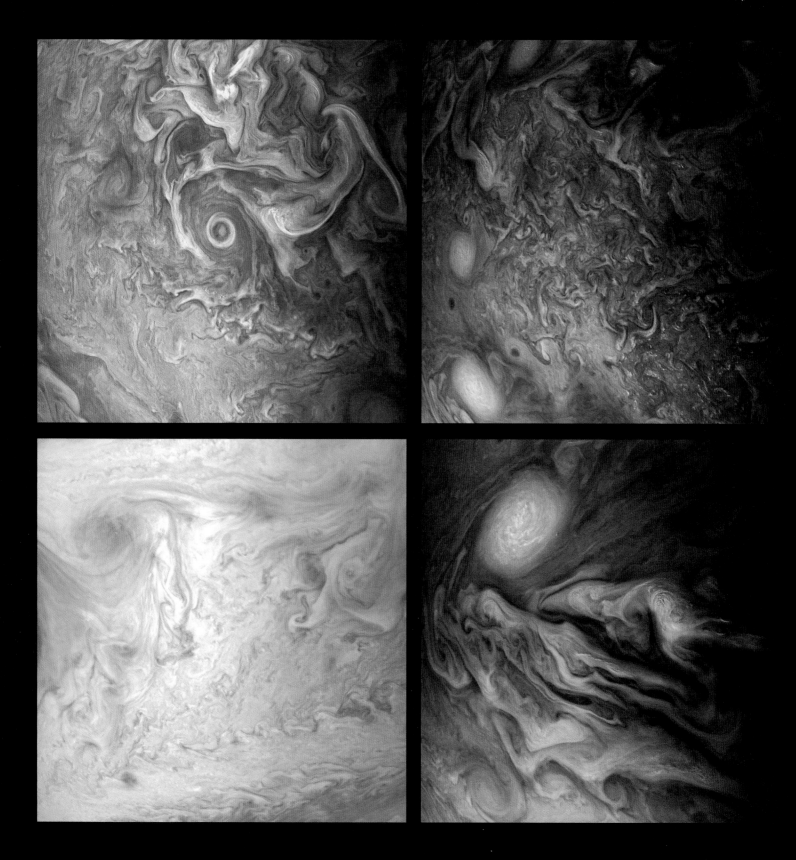

Above: Jupiter's volatile
world, as imaged in these
pictures from Juno of the
planet's cloud-covered
atmosphere.

Jupiter, the father of the skies and ruler of the heavens, had cast its hand far and wide, sculpting the precious Solar System we see today in the space of just a few million years.

Waiting in the shadows of the outer Solar System, another massive planet was lurking that we now think pulled Jupiter back from the brink. Saturn, a gas giant not as big as Jupiter, but enormous none the less, began its own slow drift inwards towards the Sun. Formed much further out than Jupiter, as Saturn began turning inwards it would have a profound effect on Jupiter's trajectory.

As the two giants drew closer together they became locked in an intimate gravitational dance, an orbital resonance that meant that for every orbit Saturn made around the Sun, Jupiter made two circuits, exerting a regular periodic gravitational influence on each other as the two giant worlds passed close by. The result of this orbital resonance between Jupiter and Saturn cleared out all the gas and dust between the two planets, creating a gap in the disc of the early Solar System. And this interaction between the two resonant planets and the gas and dust at the edge of the gap, resulted in both planets changing tack. The orbits of both Saturn and Jupiter began to move outwards, away from the Sun, preventing Jupiter from moving any further into the inner Solar System. Earth was given a reprieve and Jupiter was dragged back out into the asteroid belt once more.

This change in direction has become known as the Grand Tack Hypothesis, a nod to the action of a sailboat as it 'tacks' and changes course while it travels against the direction of the wind. Just like a sailboat, the theory suggests Jupiter 'tacked' around 1.5 au from the Sun (in the current orbit of Mars), changing direction and pulling back out to its current orbit at 5.2 au. It's a theory that's been driven by using advanced computer modelling, which helps to explain many of the inconsistencies between our Solar System and the hundreds of other planetary systems we have recently observed. It's the Grand Tack that accounts for the lack of Super Earths and explains the presence of only a handful of small rocky worlds with thinner atmospheres than those observed around other stars. The Grand Tack also allows us to explain the unusually small size of Mars, which simulations predict should be at least half the mass of the Earth, if not bigger, rather than the 0.107 Earth's mass we find it to be. It also helps explain the odd inhabitants of the asteroid belt, transformed by Jupiter's journey not just once but twice across this region of space. This is why we see half-finished planets like Ceres and Vesta left starved of building material and an asteroid belt that has a surprisingly small mass, with objects that appear to originate from both inside and outside Jupiter's orbit and have a wide range of eccentric orbits themselves.

Jupiter, the father of the skies and ruler of the heavens, had spread its hand far and wide, sculpting the precious Solar System we see today in the space of just a few million years. Before coming to rest, though, it would grant one more precious gift to the newly forming planet Earth. As Jupiter was dragged out, crossing the asteroid belt for a second time, it encountered a region full of objects rich in water and bearing minerals, and Jupiter's trajectory caused these objects to be flung back into the inner Solar System. They crashed into the young planets and moons with an intensity never seen before or since, in a period known as the Late Heavy Bombardment, it transformed the makeup of the inner Solar System, delivering a large percentage of the water Earth holds on its surface today. It's extraordinary to think that it was Jupiter that was responsible for the delivery of the most important ingredient of life, the water that provided an ideal environment for life to spark and thrive in one unbroken line for the last 4 billion years.

'Potentially, despite its destructive influence, Jupiter might have been the giver or the bringer of life to the inner Solar System.'
Leigh Fletcher, Jupiter Icy Moons Explorer mission

This story offers an extraordinary understanding of the power Jupiter has wielded over all of our lives. The more we learn about the formation of solar systems and the more we try to model a system like ours with four terrestrial water-rich planets close to the star and the gas giants further out, the more we learn that perhaps all these things had to happen. These unlikely movements and interactions between the planets were necessary to produce a system like ours and a planet like ours.

So if events like the tack of Jupiter are (as we believe them to be) extremely uncommon, the conditions for the formation of Earth-like planets, terrestrial worlds with plenty of liquid water beneath the gentle pressures of a thin blue line, must be exceedingly uncommon. It's a humbling thought to begin to comprehend that the Earth may not just be a rare rock in our Solar System but a rock that only exists because of the rare system of planets within which it evolved. All the

Top: Earth might be a rare rock that exists only because of the rare system of planets like Jupiter within which it evolved.

Top: Earth might be a rare rock that exists only because of the rare system of planets like Jupiter within which it evolved.

Bottom: While Jupiter's Great Red Spot storm continues to rage, the planet itself seems to have settled into a regular orbit in the asteroid belt.

evidence is beginning to suggest that we may be living on a tiny oasis within a vast desert, devoid of life and stretching out across the Milky Way.

Today Jupiter has settled into a regular orbit at the far side of the asteroid belt, its days of marauding through the Solar System at an end. Established in a stable orbit for almost 4 billion years, it has looked down on our planet as a distant light in the sky, unchanging while the work of evolution by natural selection has transformed the first sparks of life on Earth into the endless beautiful forms that we witness today. But although distant and seemingly detached, Jupiter's influence has been far from benign; the history of life on Earth has been interrupted by numerous cataclysms, impacts with asteroids that have fundamentally shifted life again and again. And the hand behind these disruptions? It's Jupiter that has continued to hold immense sway over our planet, exerting its influence on the asteroid belt with just the smallest of nudges – the difference between existence and extinction.

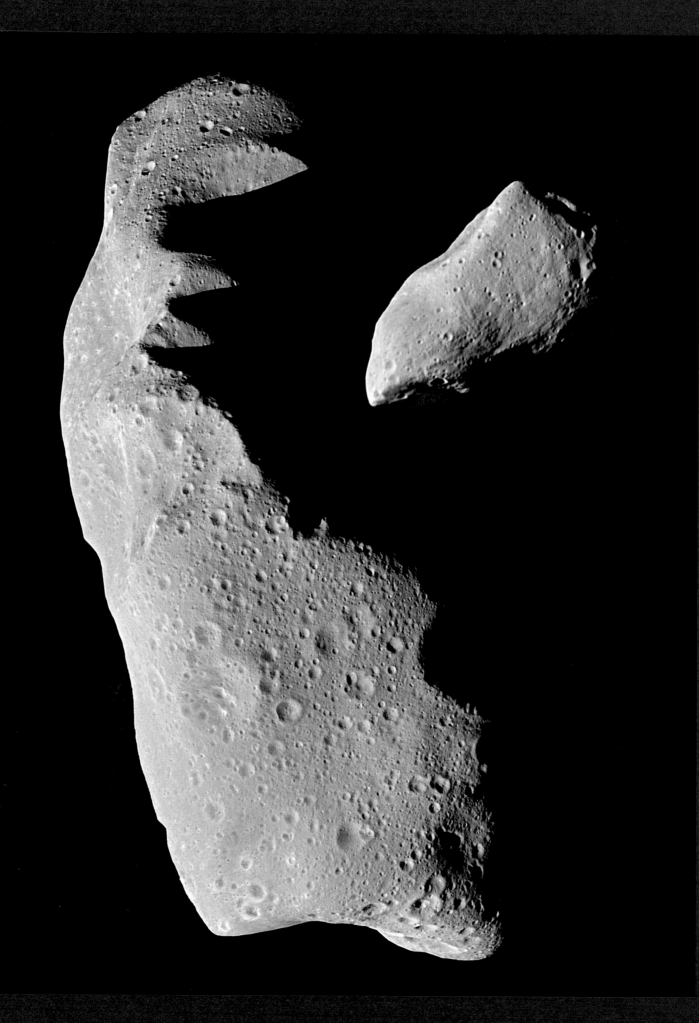

CATCHING A COMET

At 16.53 UTC on 18 October 1989 the space shuttle Atlantis took off from the Kennedy Space Centre, in Florida, carrying the Galileo spacecraft carefully packed away in its hold. Released into Earth's orbit from the cargo bay just a few hours later, the probe would begin a six-year journey to Jupiter that would utilise flybys of Earth and Venus to propel it on its 4.6-million-kilometre voyage across the Solar System. Entering orbit around the giant planet on 8 December 1995, Galileo would not only become the first spacecraft to orbit Jupiter but also the first craft to orbit any of the outer planets. It spent nearly eight years in the Jovian system and explored the giant planet and its menagerie of moons in intricate detail for the very first time, revealing a host of new discoveries. From precise measurements of Jupiter's atmosphere and observation of its giant thunderstorms, to discovering liquid water under the surface of the moon Europa and the first detailed studies of the intense volcanic activity of Io, this was a mission that opened our eyes to the beauty and diversity of the Jovian system in a level of detail that we could scarcely have imagined possible. But long before Galileo even arrived, this two-ton explorer would already have changed our view of the Solar System.

Entering the asteroid belt in August 1991, Galileo performed the first close-up flyby of an asteroid, passing within 1,500 kilometres of 951 Gaspra and returning the first detailed images of this rocky remnant back to Earth. Its next close encounter would come almost two years later when, flying past 243 Ida, the probe discovered that this asteroid had its own moon. Dactyl, as it has become known, was the first asteroid moon ever to be discovered, and the images are beautiful – a tiny 1.5-kilometre-wide object in a chaotic orbit around its parent rock.

After this unexpected encounter Galileo continued its long journey across millions of kilometres of empty space, still four years away from its ultimate rendezvous with Jupiter. But by luck rather than judgement this tiny probe happened to be travelling towards Jupiter at the precise moment it would be able to witness one of the rarest of shows our Solar System has to offer.

On the night of 24 March 1993, as Galileo sped across the Solar System, three astronomers working at the Palomar Observatory in California were attempting to spot and track what are known as Near-Earth Objects – undetected meteors that might be of potential threat to our planet. That evening, David Levy, along with Carolyn and Eugene Shoemaker, took a series of images using the 1.5-foot Schmidt Telescope. Lurking in one of them was an unusual artefact that grabbed the attention of the team. Streaking across the image was an object that, although nowhere near the Earth, was of immediate and immense interest, a comet that appeared not only to be unusually close to Jupiter but also seemed to have multiple nuclei, a fragmented series of rocks trailing across 1 million kilometres of space.

In a moment of wondrous astronomical serendipity, the team had discovered the first ever comet in orbit around another planet, a comet that had not only been captured by Jupiter's immense gravity but had also been pulled apart by it. Shoemaker–Levy 9, as it became known (because it was the ninth comet this team had found), was a comet unlike anything we'd seen before. Quickly confirmed by further studies to be orbiting Jupiter and thought to have been captured from its solar orbit back in the late 1960s or early 1970s, astronomers were able to wind back through its orbital history and calculate the precise moment the comet was ripped apart.

'Let's go for auto-sequence start. Main engine start ... and we have lift off of Atlantis and the Galileo spacecraft bound for Jupiter.'
Launching Galileo, 18 October 1989

Opposite: Asteroids Ida (left) and Gaspra (right) are two of billions of such rocky or metallic objects that orbit the Sun and could potentially collide with Earth.

Above: Galileo was released into Earth's orbit in October 1989 at the start of a six-year journey across the Solar System.

Passing just over 40,000 kilometres above Jupiter's cloud tops on 7 July 1992, we think Shoemaker–Levy 9 fell within the planet's Roche limit and this 5-kilometre comet was ripped apart by the huge tidal forces generated this close to the planet. It disintegrated into 23 pieces visible to us here on Earth, labelled fragments A to W with the biggest up to 2 kilometres in diameter. This alone was a remarkable discovery, a visual representation of Jupiter's colossal gravitational power, but if that wasn't enough, it quickly emerged from orbital calculations that the comet seemed to be on a trajectory that would ultimately take it straight through the centre of the planet. Shoemaker–Levy 9 was on a collision course with Jupiter, and as if we had scheduled it ourselves, the date of the predicted impact was July 1994 – perfect timing for the spacecraft we already had en route.

By the beginning of July 1994, Galileo was 238 million kilometres from Jupiter, on the very edge of the asteroid belt and speeding through space towards its final destination. Ahead of it comet Shoemaker–Levy 9 was entering its final few days in orbit around the planet. With the first impact expected on 16 July, the world's most powerful terrestrial telescopes were trained here, as well as some of our space observatories – from the newly fixed Hubble Space Telescope to the Ulysses spacecraft, which turned its gaze from its primary job of watching the Sun. Even

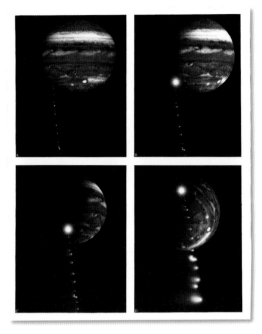

Galileo had given us a ringside seat for this rarest of phenomenon ... It transformed our understanding of the composition and atmospheric dynamics of Jupiter.

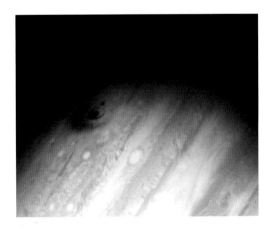

Top: Photo artwork of Shoemaker-Levy 9 impacting Jupiter, shown from multiple perspectives.

Bottom: Shoemaker-Levy 9 colliding with Jupiter, 20 July 1994, leaving a mark on the planet larger than the diameter of Earth.

the distant Voyager 2 probe, 6.6 billion kilometres away, beyond Neptune, was able to look for radio-wave emissions from the impact and employ its ultraviolet spectrometer to make distant observations. But for all of this technology peering across the Solar System, there was only one set of eyes that would be able to see the actual first impact. The comet would first hit Jupiter on the side turned away from the Earth, but Galileo was approaching on the night side of the planet, which would give us a unique view, not visible from Earth, of our first ever observation of a collision between two objects in the Solar System.

The first impact occurred at 20:13 UTC on 16 July, when fragment A entered Jupiter's southern hemisphere travelling at 60 kilometres a second, hitting the planet with an estimated force of 300 million atomic bombs. With its instruments trained on the impact, Galileo recorded the subsequent fireball which reached temperatures of over 24,000 degrees Kelvin, creating huge plumes that stretched over 3,000 kilometres high. It was a breathtaking spectacle that quickly came into view from Earth as the rapid rotation of Jupiter brought the impact sites into clear sight, revealing a huge dark spot on the surface.

Over the next six days, 21 distinct impacts were observed, as each of the fragments impacted with its own distinct set of characteristics. Fragment G was the largest of all the impacts; striking Jupiter on 18 July, it created a vast scar on the surface of the planet stretching over 12,000 kilometres across. This extraordinary spectacle continued until 22 July, when fragment W brought the show to a close, leaving just the scars of the impacts to slowly fade away over the following months.

Galileo had given us a ringside seat for this rarest of phenomena; aided by our technological eyes and ears on Earth and across the Solar System, we were able not only to witness a collision like this for the very first time, but also to harvest a vast amount of data. It transformed our understanding of the composition and atmospheric dynamics of Jupiter, allowed us to precisely measure the actual size and composition of the comet itself, but perhaps most important of all, it enabled us to study a cosmic collision first hand, revealing one of the most important dynamics of Jupiter and the power it exerts over every object and planet in the Solar System, including Earth.

Before it fragmented, Shoemaker–Levy 9 was at least 2 kilometres wide and weighed over 800 million tonnes, travelling through the Solar System in an orbit that would have taken it hurtling through the inner terrestrial planets and perhaps even close to Earth. If such an object had hit us the consequences would have been devastating, sending a vast amount of dust and debris into the air, creating a global smog that would have darkened the planet and cooled the atmosphere, leading to a catastrophic impact on all life on Earth. We will never know whether Shoemaker–Levy 9 would have posed a real danger to our planet, but we do know that Jupiter removed the threat, and such acts of planetary protection are not unusual. The impact of this comet demonstrated in spectacular fashion the role of Jupiter as a 'cosmic vacuum cleaner'. Using its vast gravitational power to capture objects like SL-9, by bringing them into its orbit and incinerating them on impact we think Jupiter acts as the grand protector, luring in objects that might otherwise pose a threat of collision with the Earth. But as we have so often seen with the story of Jupiter, no act is is ever truly benevolent; for every potentially dangerous object it shields us from, it can throw another in our direction. That's because of the influence it exerts over all the objects in the asteroid belt.

ASTEROID EVENT HORIZON

It's easy and quite natural to think of the asteroid belt as a single, homogeneous structure, a ring of rocks orbiting in an orderly fashion around the Sun, but it's nothing like that at all. In reality, each of the millions and millions of pieces of rock is following its own individual orbit, distorted and eccentric orbits that create what appears to be a chaotic cavalcade around the Sun.

But although there are complex patterns, there is also structure, and nothing has more influence on that structure than Jupiter – the silent conductor of the asteroid belt. In some way every asteroid dances to Jupiter's tune – whether it's the Trojan asteroids orbiting ahead of and behind Jupiter, or the Hilda asteroids that orbit around in three groups due to the delicate interplay between the gravitational pull of the Sun and that of Jupiter. Every asteroid is in some way caught between the two largest objects in the Solar System. In such a chaotic, complex system, collisions between asteroids are inevitable and disturbances to an orbit are commonplace.

METEOR CRATER FORMATION

Simple crater

Complex crater

All it takes is for an asteroid to line up with Jupiter and the resultant gravitational kick it receives can send an asteroid hurtling outwards, banishing it to the far edges of the Solar System or inwards towards the inner planets. It's by this mechanism that Jupiter continues to exert its influence across every corner of the Solar System.

Over the last 4 billion years, from its distant orbit 778 million kilometres away from the Sun, Jupiter has repeatedly played with the destiny of our planet and the path of life upon it. The surface of the Earth is strewn with the evidence of cataclysmic collisions that have significantly shaped the history of the planet – from the delivery of huge volumes of water almost 4.5 billion years ago, to the near-planet-destroying events of the Late Heavy Bombardment and even the formation of the Earth–Moon system. All of these momentous events can be linked to the shifting influence of Jupiter and its gravitational stranglehold on objects across the Solar System.

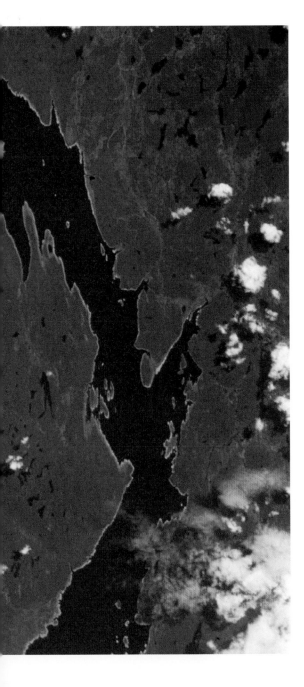

Left: A circular lake with a large island in the middle, like Lake Manicouagan, in Quebec, is a clear example of a complex impact crater on Earth.

Top: When a meteor strikes the solid surface of a planet, rock and earth spray out in all directions and a spherical crater is created. If a meteor is large enough, the rock beneath the crater will rebound, or push back up, creating a central peak in the middle of the crater.

Above: Artwork of the Chicxulub crater on the Yucatan Peninsula, Mexico, which might have caused the extinction of the dinosaurs.

Steven Desch, Icy Worlds Programme, on the
Chicxulub crater
*'If something the size of the Chicxulub impactor that
killed the dinosaurs were to happen today, it would
just be utterly catastrophic for mankind. The most
immediate thing is you would have a magnitude
13 earthquake. And then you would have tidal waves
that could happen anywhere near water, you'd have
tidal waves that are miles high propagating out for
thousands of miles in every direction. And that's
just for starters.*

*Eventually so much dust would be kicked up in to
the atmosphere and it would block sunlight, destroy
agriculture, our infrastructure would fail.'*

Over millions of years the evidence for many of these collisions has been lost or hidden under the ever-changing surface of our planet, but as we have been able to see at impact sites like the Barringer crater, Jupiter's hand is never far away from shaping events on the surface of Earth. The discovery of one such impact crater underneath the Yucatan Peninsula in Mexico back in the late 1970s has revealed evidence of an event that in a single moment changed the course of life on Earth perhaps more profoundly than any other, transforming the very possibility of our own existence.

Around 66 million years ago, in the far reaches of the asteroid belt, an object at least 15 kilometres wide was making its final orbit of the Sun. Just one asteroid among millions, this insignificant object had been following the same path for millions if not billions of years, but then something changed. Exactly what caused it to alter its path will never be known, but the most likely explanation is that a gravitational nudge from Jupiter pulled it out of its innocuous orbit and sent it on a very different trajectory, a direction that would ultimately bring it on a collision course with Earth.

Travelling through space, it would have descended through our atmosphere in just a few seconds before striking the Earth with a force equivalent to 100 million megatons of TNT. The impact generated a fireball so hot that anything within a 1,000-kilometre radius was instantly vaporised, leaving a crater 180 kilometres wide and 20 kilometres deep. Today the town of Chicxulub sits near to what we think is the centre of the impact site and lends its name to the crater that stretches out west onto the Yucatan peninsula and east into the Gulf of Mexico.

The evidence for this giant impact doesn't sit just in the partially hidden Chicxulub crater, but also in a thin layer of sediment found throughout the world that contains unusually high levels of iridium. This metal is exceptionally rare in the Earth's crust but is found in abundance within asteroids, so we believe its presence in the thin layer of sedimentary rocks around the globe dating back 66 million years ago is a direct remnant of the huge rock that struck the Earth. This geological signature is known as the K-Pg boundary, because it marks not just the impact itself but a boundary between two epochs of life on Earth, the end of the Cretaceous Period and the beginning of the Paleogene.

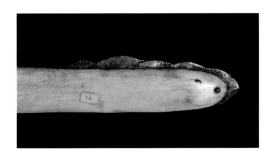

Opposite and top: Images of the Chicxulub site showing geology (top) and gravity (opposite), revealing small areas of higher gravity, caused by deep concentrations of rock. The white line indicates the coast; land is in the lower half of the frame.

Bottom and right: Meteorites have long been part of human history. The Cape York meteorite (right) was mined for metal to make tools by many generations of Inuit people in what is now Greenland.

It was an event that lasted just a few seconds, but in the aftermath of the impact millions of tonnes of sulphurous rock were thrown up into the atmosphere and during the 'nuclear winter' that followed, 75 per cent of all species were wiped from the face of the Earth, including the largest creatures ever to walk the planet.

In a global extinction event that affected every continent simultaneously, the age of the lizards was brought to an end as almost every species of dinosaur and many other species vanished. According to Steve Desch, Professor of Earth and Space Exploration at Arizona State University, it was an event we would not like to see the likes of again. 'If something the size of the Chicxulub impactor that killed the dinosaurs were to happen today, it would just be utterly catastrophic for mankind. The most immediate thing is you would have a magnitude 13 earthquake. And then you would have tidal waves that could happen anywhere near water, that are miles high propagating out for thousands of miles in every direction. And that's just for starters. Eventually so much dust would be kicked up into the atmosphere it would block sunlight, destroy agriculture, our infrastructure would fail.'

Today just traces of the dinosaurs remain. These creatures that ruled the Earth are now nothing more than a dusty history captured in the fossil record. One event, one asteroid, one orbit that was very possibly changed by the might of Jupiter – without the distant planet's intervention it's possible that those giant lizards might still be here today and the evolutionary niche that opened up for us with their removal might never have appeared.

There are many chance events, not only throughout the history of Earth, but also spanning the entire history of the Universe for 13.8 billion years, without which we wouldn't exist. There's an unbroken chain of life stretching back 4 billion years here on this planet but, having said that, there are major events that affected life, and one of them without doubt is the asteroid that wiped out the dinosaurs. And so in that sense, it is true to say that without Jupiter, we wouldn't be here. The godfather of the planets paved the way for humankind to inherit the Earth, and it has looked down on us benevolently ever since, using its immense power to keep us safe. A sleeping giant with its power undiminished, an asteroid impact like the one that cleared the way for us to evolve 65 million years ago would have catastrophic consequences for our civilisation were something similar to strike today.

To the naked eye, Jupiter is one of the brightest points of light in the night sky, and through a small telescope it is a beautiful banded world. It feels distant, eternal, disconnected from events here on Earth, but the more we've learnt about the history of the Solar System, the more we've come to understand that that is not the case. Jupiter has played an important and perhaps decisive role in the stories of all the planets, including Earth.

You see, the Solar System is just that: it's a system, complex, interconnected and interdependent. So next time you see Jupiter, just hold your gaze and think for a while about the fact that it is so much more than a point of light or a planet even. Jupiter is the great sculptor of the Solar System, the destroyer and creator of worlds.

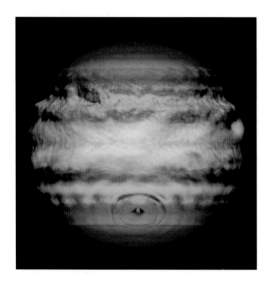

'We're not immune to having large-scale impacts on planet Earth, they have happened in the geologic past, so what might happen if this was to occur tomorrow? Well it certainly would not be very good news for the current ecosystem that we have here, but one thing I think I need to emphasise is that Earth would carry on, the planet would be very different to what we see today, but it would carry on.'
Leigh Fletcher, Jupiter Icy Moons Explorer mission

Above: Computer simulation showing the impact of a 1-km diameter comet on Jupiter's atmosphere.

Opposite: Jupiter's south pole, as seen by Juno spacecraft, a turbulent part of a wild world that is sculpting our Solar System.

SATURN

THE CELESTIAL JEWEL

ANDREW COHEN

'The most beautiful thing we can experience is the mysterious. It is the source of all true art and science. He to whom the emotion is a stranger, who can no longer pause to wonder and stand wrapped in awe, is as good as dead – his eyes are closed.'
Albert Einstein

'The Saturn system has it all: a massive gas planet, an incredible ring system, myriad worlds, icy moons with many different characteristics, and one of the largest moons of the Solar System, Titan. It's a wonderland for scientists.'
Bob Pappalardo, Cassini mission

CELESTIAL JEWEL

At any one time, 10 million tonnes of diamond 'hail' are raining down upon Saturn's surface.

Saturn is a gas giant, alive with storms of unimaginable ferocity, a world of alien weather where diamonds rain down from the sky and bubbling vortexes whip around the planet so fast they catch up with their own tails. And glistening in the distant light of the Sun is a 500,000-mile ring of almost pure frozen water, broken into millions of individual pieces. This structure is unique in our Solar System, sculpted by the delicate hand of gravity. Yet for all of their iconic beauty, the rings are just a fleeting shadow, an echo of Saturn's past and an ephemeral structure that has adorned the planet for a brief moment in its history and will soon disappear again.

Saturn is a place of many worlds, encircled by at least 62 moons. This menagerie of satellites boasts worlds of astonishing variety: Titan – bigger than the planet Mercury – is the only moon in the Solar System to have its own substantial atmosphere, while Mimas is the smallest body in the Solar System. Perhaps most astonishing of all, though, is Enceladus – an ice moon with an ocean of liquid water inside that seems to contain all the building blocks of life.

Composed of 96 per cent hydrogen and 3 per cent helium, Saturn has a diameter at its equator almost ten times the size of the Earth – at 120,536 kilometres. Its volume would hold 800 Earths. However, with its mass made up predominantly of gas, the planet's density is remarkably low. As the only planet in the Solar System less dense than water, if you could find a bathtub big enough to hold it, Saturn would literally float in it.

Beneath the frosty minus 178 degrees Celsius cloud tops, the pressure rapidly increases under the huge weight of Saturn's dense atmosphere. The hydrogen gas of the upper layers turns at first into liquid, and then, hundreds of kilometres further below the planet's cloud tops, into slushy metallic oceans of hydrogen.

We can only make an educated guess about the structures that lie within Saturn's unreachable core. With temperatures at its centre exceeding 11,000 degrees Celsius, Saturn generates far more heat from within itself than it receives from a sun that sits on average 1.4 billion kilometres away, meaning that it must be generating its own heat. With such extreme temperatures and pressures, the structural makeup of the planet is uncertain and hotly debated. According to planetary scientist and physicist Jonathan Lunine, 'if you go down, down, down, through the hydrogen, through the helium rain, to this core of rocky and icy material, it's not rock and ice. It's some sort of atomic crystal where the elements of rocks and ice, the silicon, the oxygen, the magnesium are all together in a structure that doesn't exist in the rocks on the Earth. Now in the case of Jupiter, it might actually be warm enough that the core is a liquid, that it's molten. In the case of Saturn, it's probably a solid core, but definitely an exotic one.'

Undoubtedly there is a dense core at the heart of Saturn, one that is almost certainly composed of iron-nickel alloy and rocky silicon-based compounds, but exactly how this material behaves at such extremes remains a mystery.

Opposite: Saturn sits stately in our Solar System, surrounded by its iconic icy rings.

Above: The Cassini mission sent back revelatory images of the planet, such as this one, showing clouds in Saturn's northern hemisphere.

'Scientists thought of the Saturn system as recording the history of the Solar System, but didn't realise it would be a very active place. When we got there we found that it's very dynamic, that the planet itself changes with the seasons.'
Bob Pappalardo, Cassini mission

From its surface all the way down to its centre, Saturn is a world of mystery. This gas giant, along with its sister planet, Jupiter, is worlds apart in character from Mercury, Venus, Earth and Mars – the terrestrial planets of the inner Solar System. It is hard to imagine how they could all have been created from the same ingredients, the singular cloud of gas and dust that gave birth to every planet in our Solar System. But if we go back far enough in time, we begin to uncover a history with a surprisingly familiar origin. The beginning of Saturn's story is far more recognisable than its present-day character would suggest, because at the heart of all of the ringed planet's beauty lies a lost world: a primordial one that disappeared from view billions of years ago and yet was the seed for everything that Saturn was to become. Without this early world there would have been no gas giant, no rings of ice and no hope of an orbiting oasis of life in the far distant reaches of our Solar System.

Below: The ice line divides a solar system into two zones based on temperature; one warm enough for liquids, the other frozen solid.

THE ICE LINE

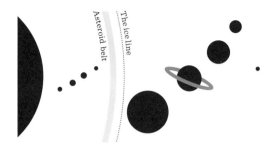

Asteroid belt

The ice line

The story of Saturn, like that of all the planets, began around 4.5 billion years ago. In the light of the awakening Sun, a vast cloud of gas and dust swirled around our newly formed star, and that protoplanetary disc of gas and dust clumped and coalesced until it slowly began to form into a collection of embryonic worlds that would go on to become the four rocky planets sitting closest to the Sun. Millions of kilometres further out, in the cold distant reaches of the Solar System, a similar story was playing out, but this tale of planetary formation would follow a very different path. That's because Saturn was moulded beyond a boundary called the ice line (sometimes also called the snow line), which is a frontier that marks the divide between the terrestrial planets and the gas giants. Respective proximity to this line would lead the evolution of Saturn and that of its neighbouring planet, Jupiter, to play out very differently.

The ice line, as you might expect from the name, is a way of dividing a solar system into two broad zones based on temperature. Inside the ice line the strength of a star is strong enough to keep things pretty warm, but step onto the other side of the line and you have crossed a boundary where it becomes cold enough for volatile compounds like water, methane, ammonia and carbon dioxide to condense and freeze solid. (Each of these compounds does, of course, have its own precise ice line, due to the different specific freezing points of each individual compound, but beyond a certain point the boundary begins to incorporate the freezing points of a large number of compounds.)

Today the ice line is thought to be around 5 au – that's around 750 million kilometres – from the Sun, which means it lies well beyond the asteroid belt, at a distance just before the orbit of Jupiter. But in the infant days of the Solar System that distance would have been much less. With a weak young Sun shrouded in an opaque cloud of dust and gas, the early ice line would have fallen about 2.7 au from the Sun, smack bang in the middle of where we find the asteroid belt today. We know this because we have been able to explore the makeup of the asteroids on either side of this line. Beyond the 2.7 au line, we've discovered the asteroids to be icy objects with large amounts of water ice locked away within them, while on the inside of this line we find a very different class of asteroid, dry and largely devoid of water. It seems these bodies formed in a region of space that was simply too warm to allow substantial amounts of ice to exist.

Why does the ice line matter in the story of Saturn's evolution? In the early Solar System, as the protoplanetary disc swirled around the young Sun, the ice line not only defined an important boundary in terms of temperature, but it also created a dramatic difference in the abundance of material available to build a planet on either side of the line. Beyond the ice line the lower temperatures meant that all of those frozen volatile compounds were swirling around in the cloud of gas as solid grains, building blocks that provided vast amounts of additional material to feed into the process of planetary accretion, and the impact of this simple physical difference was profound. On our side of the line we find only rocky planets like Earth, Venus and Mars, while on the other side of the line there is not a single planet that we recognise as terrestrial, just the gas giants of Saturn and Jupiter and the ice giants of Uranus and Neptune beyond. To understand the reason for this stark delineation we need to look back to a critical time in Saturn's history, to the moment when the paths of the planets either side of this ice line began to diverge.

Left: This composite image shows Cassini in orbit around Saturn.

Above: The origin of Saturn
lies in the accumulation
of rock and ice to form
one of the biggest worlds
in the Solar System.

Saturn began its life not as a beautiful jewel but as a tiny, ugly, ragged world. Just like its terrestrial cousins, the origins of this giant gas planet began with the clumping together of tiny specks of gas and dust that slowly grew into bigger and bigger clusters. As this embryonic new world tumbled wildly through space, days would have lasted minutes as the distant Sun lit its twisted surface, rising and falling in chaotic rhythm. Eventually, as it grew, this ragged rock began to transform itself, taking the new shape of a spherical world as it went from planetesimal to protoplanet.

On the warm side of the ice line, the growth of the terrestrial planets came to an abrupt halt. As these planets had secured stable orbits and began clearing their way through the protoplanetary disc, the supply of raw ingredients ran dry and so their growth simply stopped, starting with the runt of the litter, Mars, followed by Venus and then our planet Earth, the largest of the rocky planets. But on the other side of the ice line it was a very different story. With no shortage of building material, the growth of the protoplanetary Saturn was able to progress at a rapid rate. With abundant amounts of rock and ice the young Saturn continued to grow, quickly dwarfing any of the terrestrial planets that were forming in the far distant glow of the warming Sun.

Over a relatively short period of hundreds of thousands of years, this plentiful supply of material smashed and clumped and grew until, through a series of planetary-scale collisions, Saturn became the largest object for millions of miles in any direction. The bigger it got, the stronger its gravitational field grew, and so it could attract ever more material into its grasp. Eventually the violence of its early life allowed Saturn to grow into one of the biggest worlds of rock and ice the Solar System had ever seen.

What did this world look like? We simply do not know. The detail of how any of the planets formed in the Solar System is only sparingly understood; these are events that happened 4.5 billion years ago and it's very much still cutting-edge scientific research trying to piece together the barest of timelines, let alone the intricate details of these distant events. Much of what we do understand of the evolution of our own Solar System and the planets within it has been supported by our recent ability to study worlds outside it, in neighbouring systems. The study of exoplanets has allowed us for the first time to peer into the evolution of other star systems and witness the formation of hundreds of different planets rather than the eight in our own backyard.

Much of this evidence supports the idea that Saturn was once a world of rock and ice, a world that grew to perhaps 10 or even 20 times the mass of the Earth as it gorged itself on the available matter. This was a world we would at least have vaguely recognised, a place it is almost possible to imagine, a world that you could stand on and survey, but it was a world that wouldn't last.

The fact that Saturn grew to such a size had significance beyond just scale. Its size pushed the planet towards a very different evolution, changing its relationship with the environment and ultimately sending it on a completely new path. It's because of its immense mass and its place in the Solar System that Saturn wouldn't remain a rocky world for long.

Top: Saturn's atmosphere; a storm is brewing in the north of the planet, with high clouds and haze beneath it, coloured blue and white. The rings cast a thick shadow.

Bottom: Studying star clusters such as this one, NGC 2362, has helped scientists to understand how gas giants like Saturn form so rapidly.

SATURN'S FORMATION

'Saturn and Jupiter are gas giants. This means they are mostly comprised of hydrogen and helium, which are the two lightest gases in the Universe, and what stars are made out of. Basically, Jupiter and Saturn are failed stars. If they had just grown about ten times bigger, they might have exploded into stars and we'd have a double star system in our Solar System. We'd have two suns in the sky.

So how did Saturn form? Well, all the planets formed, we think, in basically the same way. Originally, there was a big cloud of gases and materials, actually the whole Solar System is a second- or third-generation conglomeration of materials that were originally in other stars, and those stars exploded and this is like a reprocessing of the stuff that came out of those stars.

So you have a nebula of materials and you get a core through instabilities and just randomness, which becomes the centre and will become the Sun, eventually. But in the beginning, the core forms, a little condensation, a little bit of density increase, and then gravity makes the cloud start rotating.

Then you get a rotating disc, eventually, and inside that disc you'll find little whirls just happening, instabilities occurring in certain places. These little places condense more material. Then that forms its own gravity and it starts running away. So in the end you get these planets forming by what we call the core accretion model, where you first form the dusty-type things that can glue together to themselves, and then you form a little planetesimal and it grows to the size of

our Moon, after a couple of hundred thousands of millions of years. Eventually, it gets so big that now it can grab onto the rest of that cloud, which is formed of these condensable light gases, water and other types of materials, so then it can get an atmosphere, a huge atmosphere.

The actual formation mechanism is still up for debate, but models show this can happen just in a few million years; once you get that instability happening and there is enough material in the core. It looks like the cores of Saturn and Jupiter caused those planets to grow to many hundreds of Earth masses. So that suggests they are predominantly hydrogen–helium. Jupiter grew to about a thousand Earth masses and Saturn just slightly less – about 700 or so. As this is happening, the

Sun's turning on, so the Sun is also condensing, and it is then reaching the pressure and temperature where it can turn on and put out light.

That light, when it comes out, spreads everything out, so the hydrogen–helium that was there when the Earth was still forming gets exploded out and it kind of flies out to the orbits of Jupiter and Saturn, where that water vapour can condense out. It's so far from the Sun, now it can condense out into ice, which changes the way it moves, and now you can more easily grab onto it.'

Kevin Baines, Cassini mission

SEEKING SATURN

'Nothing considerable in
[scientific knowledge] can be
obtained without secrecy.'
Robert Hooke to the Royal Society

The history of the exploration of Saturn is littered with many of the greatest names in the field of science. Galileo Galilei was the first to peer at her through a telescope in 1610, mistaking the faint outline of the rings for two moons on either side of the planet – he famously reported the planet appeared to have ears. Two years later, however, when he looked up again, the peculiar structures on either side of the planet had disappeared. We now know this is because the rings were facing the Earth edge-on at the time of observation, but for Galileo this was a strange and mysterious anomaly. It would take almost 50 years and the more powerful telescope technology of Christiaan Huygens to solve the problem. Using a 50 mm refracting telescope, Huygens was the first to spot a moon of Saturn, publishing his discovery of Titan in 1656. At the same time his observations led him to begin formulating a theory to explain the strange 'ears' that Galileo had first seen – not moons, but a ring around the planet. Uncertain that he had enough proof to convince the scientific community of his hypothesis, Huygens did what was common at the time: he published his thinking not as a preliminary hypothesis or even as a line of

Above: View of the Earth, taken by Cassini while in orbit around Saturn, from a distance of approximately 1.44 billion km.

Earth

investigation, but instead, in order to mark the territory and protect his reputation, he announced his discovery in the form of an anagram:

aaaaaaacccccdeeeeeghiiiiiiilllllmmnnnnnnnnnnoooopppqrrsttttttuuuuu

For two years the anagram remained unbroken, protecting Huygens's claim to be the first to identify the rings while also giving him enough time to build the evidence to make the hypothesis defendable. It wasn't until 1658 that Huygens revealed the meaning of the anagram to be a Latin sentence: *'annulo cingitur tenui, plano, nusquam cohaerente ad eclipticam inclinato'*, which means: 'it is surrounded by a ring, thin and flat, never touching, oblique in relation to the ecliptic'.

With the rearrangement of 62 letters, one of the most beautiful structures in the Solar System, the rings of Saturn, had been revealed to the world, if not yet understood. While Giovanni Cassini and William Herschel went on to discover more moons of Saturn and to detail structures within the rings (including the 4,800-kilometre gap between the A- and B-ring that would later become known as the Cassini Division), we still imagined the rings as one great disc, a flat solid circle of material encircling the planet.

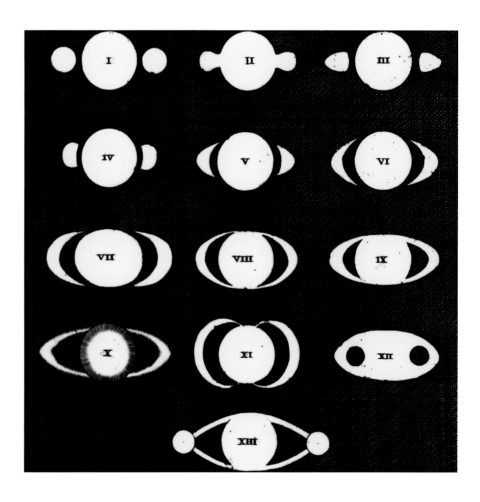

Top, bottom and right:
Scientists as far back as the seventeenth century have been fascinated by Saturn, studying their phases (left) their rings.

Pioneer 11 Mission Timeline

Time	Event
1979-08-29	
	Encounter with Saturnian system.
06:06:10	*Iapetus flyby at 1,032,535 km.*
11:53:33	*Phoebe flyby at 13,713,574 km.*
1979-08-31	
12:32:33	*Hyperion flyby at 666,153 km.*
1979-09-01	
14:26:56	*Descending ring plane crossing.*
14:50:55	*Epimetheus flyby at 6,676 km.*
15:06:32	*Atlas flyby at 45,960 km.*
15:59:30	*Dione flyby at 291,556 km.*
16:26:28	*Mimas flyby at 104,263 km.*
16:29:34	*Saturn closest approach at 2,059 km.*
16:35:00	*Saturn occultation entry.*
16:35:57	*Saturn shadow entry.*
16:51:11	*Janus flyby at 228,988 km.*
17:53:32	*Saturn occultation exit.*
17:54:47	*Saturn shadow exit.*
18:21:59	*Ascending ring plane crossing.*
18:25:34	*Tethys flyby at 329,197 km.*
18:30:14	*Enceladus flyby at 222,027 km.*
20:04:13	*Calypso flyby at 109,916 km.*
22:15:27	*Rhea flyby at 345,303 km.*
1979-09-02	
18:00:33	*Titan flyby at 362,962 km.*

It was Pierre-Simon Laplace, the 'French Newton', who was the first to suggest in 1787 that such a singular structure around a planet would be unstable and instead was more likely to be comprised of a number of solid ringlets. By the mid-nineteenth century it was the turn of the great mind of James Clerk Maxwell to investigate the rings. Unpicking centuries of conjecture around the nature of the rings, he demonstrated for the first time through a series of mathematical proofs that any solid structure, be it single ring or ringlets, would be unstable and so unable to exist around the distant planet. Therefore, Maxwell concluded, the rings must be composed of a huge number of small particles all orbiting the planet independently.

In his words: 'Every particle of the ring is now to be regarded as a satellite of Saturn, disturbed by the attraction of a ring of satellites at the same mean distance from the planet, each of which, however, is subject to slight displacements. The mutual action of the parts of the ring will be so small compared with the attraction of the planet, that no part of the ring can ever cease to move round Saturn as a satellite.'

With Maxwell's publication, *On the Stability of the Motion of Saturn's Rings*, in 1859, our modern understanding of this distinctive feature of Saturn had begun, but it would be another 120 years before we'd get the chance to explore 'the most remarkable bodies in the heavens', as Maxwell described them, up close.

At 14.26 Universal Time on 1 September 1979, after a six-year journey from Cape Canaveral across over a billion kilometres of space, the Pioneer 11 spacecraft took us within touching distance of the rings of Saturn for the very first time. As our first explorer to cross the boundary of the rings, Pioneer flew across the ring plane beyond the outer ring, enabling it to photograph the Saturnian system in extraordinary detail, sending back a series of images that revealed the ringed planet in greater resolution than we could ever have seen from Earth-bound observations. On its way through the ring plane, Pioneer was able to take the first close-up measurements of the rings, and in doing so discovered not only a new ring system, the F-ring, but also two new moons of Saturn. In fact, the spacecraft flew so close to the as-then-unknown moon, passing at a distance of just 6,676 kilometres, it was lucky the mission didn't abruptly end with a massive collision. We now know Pioneer had encountered either Epimetheus or Janus, but we're still not sure which one of these two similar-sized moons it was, as both sit in the same orbit around Saturn.

Travelling at a speed of 114,100 km/h, Pioneer passed within 21,000 kilometres of Saturn's cloud tops, making its closest encounter at 16.29 UST on that September day, before speeding off beyond the rings and onwards on its journey to the outer solar system. As well as taking 440 images of the Saturnian system, Pioneer measured the characteristics of Saturn directly. An array of sensors were able to confirm that this giant planet was a bitterly cold world with an average overall temperature of minus 180 degrees Celsius, and a composition made up almost entirely of liquid hydrogen. But beyond the basic data, perhaps Pioneer's greatest achievement was to plot a course through the Saturnian system and make its way safely through the ring system while returning data back to Earth.

Opposite top left: Pioneer 11's flyby revealed a new ring system around Saturn, the F-ring.

Opposite top right: The Cassini mission sent back images of several of Saturn's moons, including Epimetheus (top) and Janus (bottom), revealing their cratered surfaces.

Opposite bottom: Pioneer 11 measured the heat radiation from Saturn's interior and that of its planet-sized moon, Titan, pictured here.

Top: Pandora and Prometheus, the shepherding moons of the F-ring, orbit inside and outside the thin ring (second image) and the pull of their respective gravities causes constant change in the ring's shape (first image).

Bottom: Saturn's rings have been named in order of their discovery, some as letters, some as divisions. The various flybys of Saturn have enabled scientists to have an ever more detailed idea of the makeup of this planet's unique ring system.

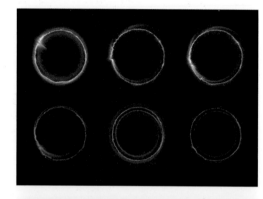

Now the stage was set for the two Voyager probes, already well on their way across space, speeding towards the ringed planet equipped with an even more advanced array of technology, ready to probe even deeper into Saturn's mysteries.

The rings were named essentially in order of discovery. In terms of the main rings, it starts at the outermost, which is the A-ring, and then goes into B-ring, where most of the mass is. And there's a faint C-ring, and an even fainter D-ring inside that. Then the E-ring was discovered around 1980, when the rings as viewed from the Earth were edge on. That's the one that's clearly associated with Enceladus. Then Pioneer 11 discovered the F-ring, and then Voyager discovered the G-ring. And the G-ring, we know, is associated with a moon called Aegaeon, which is the source material for the ring. At that stage we stopped using the letters as a nomenclature. So there are other individual rings that have been named, but they tend to be rings or ringlets, that are in the main ring system itself.
Carl Murray, Cassini mission, on the organisation of Saturn's rings

Just over a year after Pioneer had exited the Saturnian system, early in November 1980, Voyager 1 swept in with its high-definition cameras primed and ready for action. As it sped past Saturn at 17 km/s it was able to produce a wealth of new data about the planet, rings and moons. One of its key discoveries was the identification of three new moons: Prometheus, Pandora and Atlas. The mission revealed detailed relationships between these moons and the ring system within which they sat, which also confirmed a long-standing theory about the structure of the rings. At least the first two of these moons are 'shepherding moons', satellites that exist within the narrowest rings of Saturn, and play an essential role in keeping the delicate structures in place by their gravitational interactions with the billions of particles within the rings. With its wide- and narrow-angle cameras, Voyager 1 was able to take a series of pictures showing in incredible detail Prometheus acting as a shepherding moon to the F-ring and Atlas as a shepherd to the A-ring.

D-ring

C-ring

B-ring

This first Voyager craft was also sent by the team at NASA's Jet Propulsion Laboratory to investigate one of Saturn's most intriguing moons up close for the very first time. Titan had already been identified as potentially the only moon in the Solar System with a thick Earth-like atmosphere and Voyager was set on a course that enabled it to pass by within just 4,000 miles of its surface. As it flew on the dark side of the moon, looking back at Titan's atmosphere through the haloed light of the Sun, it was able to accurately measure for the first time the density, composition and temperature of the atmosphere, and also obtain a precise measurement of Titan's mass. This mysterious moon, the largest of all Saturn's satellites, was found to have an atmosphere almost entirely composed of nitrogen, just like the Earth, with a surface pressure just 1.6 times that of our planet. Voyager opened our eyes to this curious world for the first time, a world we would come to explore in even more detail with the Cassini-Huygens mission, which would touch down on the surface of Titan.

As well as exploring the rings and moon systems of Saturn, Voyager 1 was able to deploy its array of sensors and cameras on the planet itself. With its closest approach on the 12 November 1980, Voyager came within 77,000 miles of the Saturnian cloud tops and was able to measure powerful winds travelling at 1,100 mph, whipping around the planet's equator and aurora. But most intriguing of all was the discovery that Saturn's upper atmosphere was composed of only 7 per cent helium while the rest was mostly made up of hydrogen. Scientists had expected Saturn to have the same atmospheric helium levels as Jupiter and the Sun, at 11 per cent, so this much lower level suggested something intriguing was going on deep within Saturn's internal structures. Where was the missing helium? The lower levels in Saturn's upper atmosphere suggested that the heavier helium might be slowly sinking through Saturn's hydrogen; but it would take the arrival of Voyager 1's sister craft nine months later to enable scientists to further probe the inner workings of Saturn, and to begin to reveal the fate of the rocky ice world that had been lost inside this planet billions of years earlier.

'Voyager had romance, it had just such a deep sense of human longing. It still has that iconic stature in our culture and there will never be another mission like it. There can't. We'll never be that innocent again.'
Carolyn Porco, Cassini mission

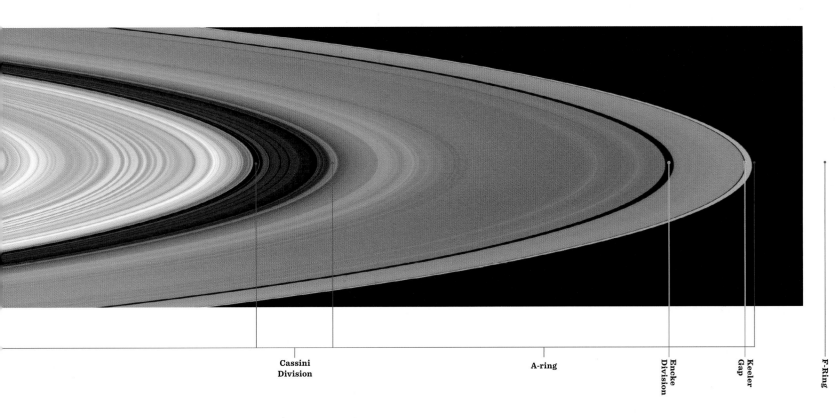

Cassini Division   A-ring   Encke Division   Keeler Gap   F-Ring

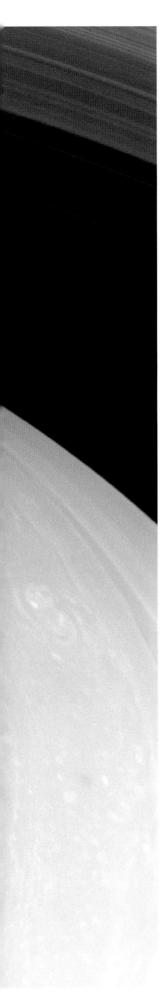

Left: This infrared image of the north pole of Saturn, captured by Cassini, shows the hexagonal cloud that is formed by the planet's polar jet stream.

While passing behind Saturn, and so shielded from the view of the Earth, Voyager 2 used its radio science system to probe Saturn's upper atmosphere, gathering information on its atmospheric temperature and density. It was an ingenious dual use of its telecommunications system that allowed Voyager to create a temperature profile of the planet that provided the first direct evidence of something that scientists had long suspected: at the uppermost levels of its atmosphere Saturn's temperature was 70 K (minus 203°C), while at the deepest levels (around 120 kilopascals) the temperature increased to 143 K (minus 130°C). Saturn, it seemed, was emanating far more heat than it was receiving from the Sun. This was a planet with a mystery at its heart, but Voyager was about to gain an even more beguiling insight into Saturn's inner workings.

In August 1981, as the Voyager 2 probe passed over the north pole of the planet, it took a series of images with its two camera imaging systems that would reveal an extraordinary secret sitting on Saturn's crown. Stitching together the images and analysing the data from this distant probe, David Godfrey and a team from the National Optical Astronomy Observatories in Tucson, Arizona, were able to piece together a startling image of a storm system unlike anything ever seen, anywhere in the Solar System. First published in 1988 , seven years after Voyager 2 had left the Saturnian system, the patchwork image revealed a puzzling hexagon-shaped weather pattern sitting around the planet's northern pole. Each of the six sides of this vast storm system is around 14,500 kilometres in length, 29,000 kilometres across and at least 300 kilometres high. It's so vast the Earth could fit into it at least four times over, and it's not just the scale of it that's breathtaking: swirling inside it are winds whipping around at speeds of over 320km/h and the whole hexagonal storm is completely rotating every 10 hours and 39 minutes. This is a storm on an unimaginable scale, and 30 years later it still rages on, but despite multiple hypotheses, the atmospheric dynamics behind its shape remain a mystery.

> '*As the seasons changed and spring arrived in the northern hemisphere towards the second half of the mission, we got the chance to look at the hexagon that surrounds Saturn's pole. If you flattened it out, it would be about two Earths wide. Saturn itself is about 10 Earths wide. So this is a major feature and it's basically a jet stream ... Every time we posted a picture of the hexagon on our website, the hits went through the roof, because I think people were so amazed that you could have something in an atmosphere that had straight sides to it.*'
> Carolyn Porco, Cassini mission

Right: These real-colour images of the north-polar region of Saturn show that the colours change through its seasons, but not the core, which stays blue.

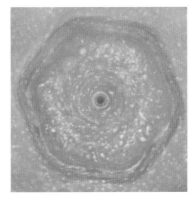
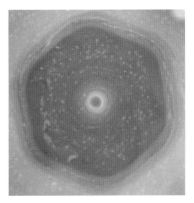

THE AIR
UP THERE

From our vantage point here on Earth it looks as if our atmosphere stretches on and out forever, and yet (by the most widely held definition) it takes a journey of just 100 kilometres to officially make your way through the atmosphere and into space. Of course, atmospheres don't just stop dead; the Earth's atmosphere gradually recedes with increasing altitude, becoming progressively thinner until, by the time you get 62 miles up, you'll have 99.999997 per cent of the atmosphere below you. This essentially arbitrary line between Earth and space is known as the Karman line. It's a line that was calculated by the Hungarian-American engineer and physicist Theodore von Kármán to be the altitude at which the air becomes simply too thin to support aeronautical flight, where, as he wrote, 'aerodynamics stops and astronautics begin'. Cross this line and you have officially gone from being a pilot to an astronaut, and you get to join a club that still only contains 561 members at the time of writing. The number of humans who have made it over the line is still so small because travelling the short hop into space might be slight in distance but it is large in terms of the power and technology needed to lift anyone or anything that far from the Earth's surface. Gravity binds us tight to our little rock, and despite over 60 years of space travel it means almost all of us are still very much rooted to the ground.

The reason we find it so tough to leave the Earth is because the mass of the planet, all 5,973,600,000,000,000,000,000,000 kilograms of it, which creates a gravitational pull so strong it requires an escape velocity of around 11.186 km/s to make it off the Earth and into space. That's a lot of speed to counteract the 9.8 m/s^2 of gravitational force that is continually tugging us down. Gravity keeps our species firmly locked to the surface of the planet, but it also keeps a tight grip on everything else on the surface of the Earth – be it solid, liquid or gas – including the 5.5 quadrillion tonnes of nitrogen, oxygen, carbon dioxide and other trace gases that together make up the Earth's atmosphere. In the very simplest terms, the size of the atmosphere of any planet is intricately linked with the mass of that planet. The bigger the mass of a planet, the greater the attraction of gases to its surface. But that's not the end of the story; the ability of a planet to retain an atmosphere is also dependent on a multitude of other factors, including the composition of the atmosphere, its temperature, proximity to the Sun, geological activity, the ability of the planet to protect itself from the ravages of the solar wind, and finally the impact of life itself, which has seen our atmosphere gradually pumped full of oxygen. It's the intricate balance of these factors that dictates the endurance and composition of the gases around a planet – each one a unique atmospheric fingerprint.

Fast-forward through the history of the Solar System and we see a familiar theme again and again around the inner rocky planets. Each of the four planets grabbed onto a thin atmosphere, but then the individual characteristics of each planet set in motion its own individual evolution. Mercury, the smallest of the planets and closest to the Sun, lost its atmosphere long ago with only the merest wisps remaining today. Mars, as we have seen in Chapter 2, was just too small to hold on to much of its atmosphere, and so even though it is the furthest away from the Sun today it is surrounded by a thin line just 50 miles deep. Next there is Earth, sitting in the Goldilocks glow of the Sun; we have clung on to our 62 miles of atmosphere in one form or another for billions of years, allowing life to thrive underneath its warming embrace. Venus, on the other hand, despite starting with

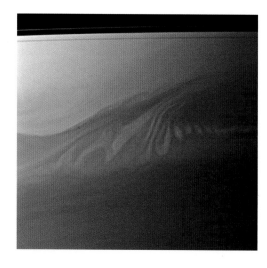

Above: Cassini imaged in fine detail the cloud formations in Saturn's atmosphere, swirling above its surface.

Opposite: The clouds off the Chilean coast on Earth show a unique pattern called a 'von Karman vortex street', which is vital in cooling the Earth's surface, keeping it at a habitable temperature.

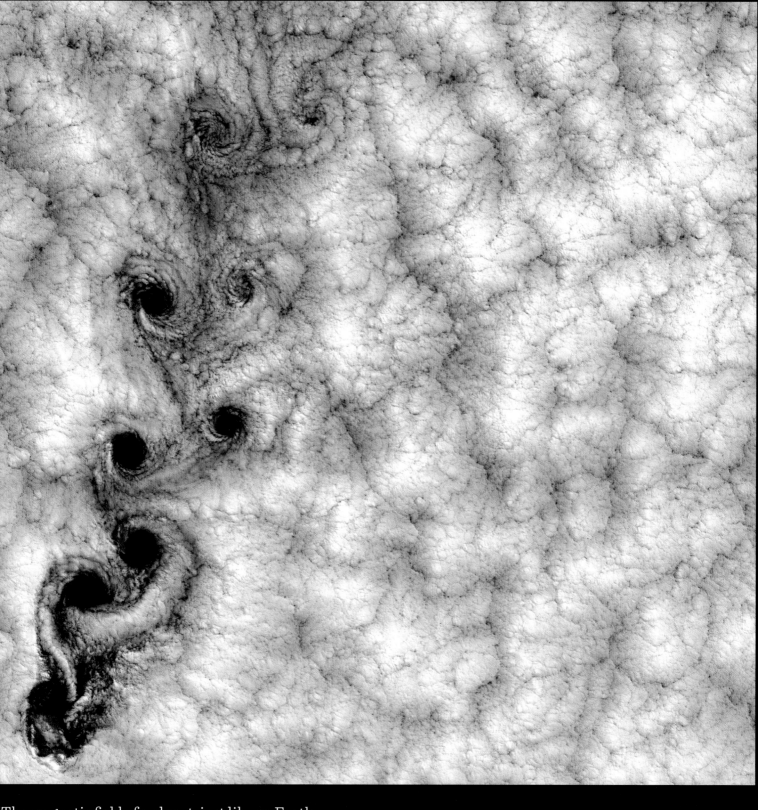

'The magnetic field of a planet, just like on Earth,
controls the behaviour of the environment around that
planet, and protects it from the effects of the solar wind.'
Michele Dougherty, Cassini mission

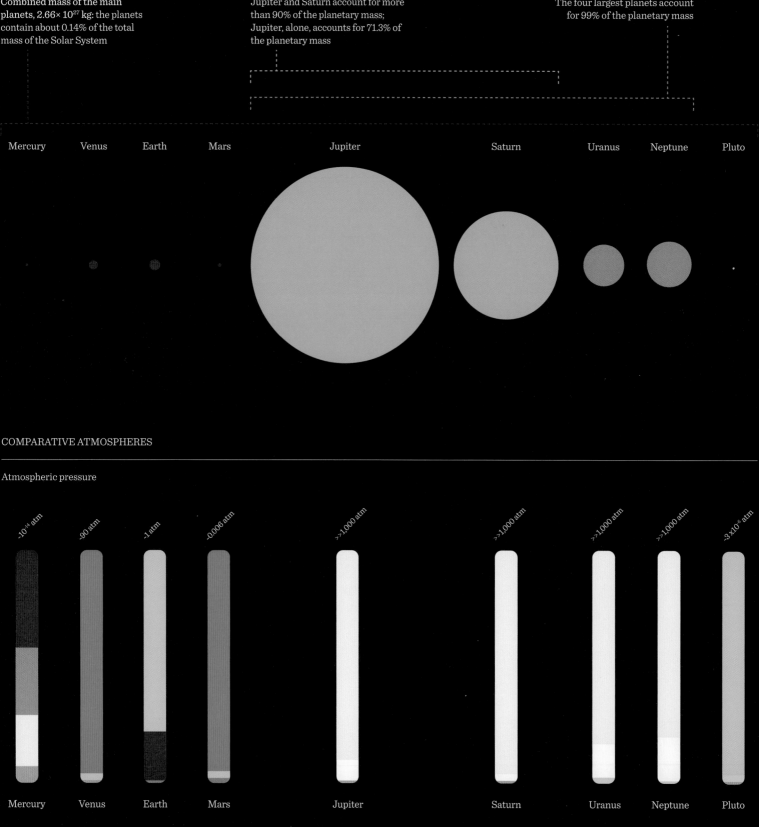

Combined mass of the main planets, 2.66×10^{27} kg: the planets contain about 0.14% of the total mass of the Solar System

Jupiter and Saturn account for more than 90% of the planetary mass; Jupiter, alone, accounts for 71.3% of the planetary mass

The four largest planets account for 99% of the planetary mass

Mercury Venus Earth Mars Jupiter Saturn Uranus Neptune Pluto

COMPARATIVE ATMOSPHERES

Atmospheric pressure

~10^{-14} atm ~90 atm ~1 atm ~0.006 atm >>1,000 atm >>1,000 atm >>1,000 atm >>1,000 atm ~3×10^{-6} atm

Mercury Venus Earth Mars Jupiter Saturn Uranus Neptune Pluto

Oxygen – O_2 Hydrogen – H_2 Nitrogen – N_2 Helium– He   Other gases

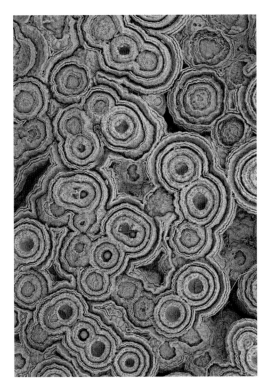

an atmosphere likely very similar to our own, has now choked on the 155-mile deep toxic atmosphere that blankets its surface today. Each of the four rocky worlds has seen its atmosphere ebb and flow over time, but no matter what twists and turns the story of each planet may have taken, there has always and will always be a limit to the size that any atmosphere can grow around these relatively small lumps of rock.

Before we return to Saturn to understand how its atmosphere grew in a very different way to those of the rocky planets, we should take a swift look through the history of the Earth's atmosphere in more detail to establish how it began and evolved. Using a combination of geology, climatology and paleontology, scientists have been able to piece together an immensely detailed picture of the Earth's atmospheric evolution.

Far back in the Earth's history, a very different type of atmosphere to the one we see today surrounded the primordial planet. The newly formed Earth had grabbed onto gases from the billowing vapours of the solar nebula, the cloud of gas and dust that the planet had formed from, consisting almost entirely of hydrogen and simple hydrogen-based molecules such as water (H_2O), methane (CH_4) and ammonia (NH_3). This primary atmosphere would have been entirely created by the capture of the nebula's gases by the gravitational attraction of the planet. Slowly, over time, the composition of the atmosphere changed from a hydrogen-based one to one rich in nitrogen and carbon dioxide. We think this shift to the secondary atmosphere was at first triggered by the giant impact of the planetesimal Theia on the early Earth, and then driven by a combination of gases produced during the Late Heavy Bombardment, while asteroids rained down on our planet for millions of years and the geological activity within the Earth sent vast quantities of gas upwards through all-encompassing periods of volcanism. By the time first life appeared on the planet, the nitrogen-rich atmosphere was stable, but gradually, as life took hold, a new component would begin to appear in the Earth's atmosphere.

Two-and-a-half billion years ago, photosynthetic life had become so prevalent it triggered a tipping point known as the great oxygenation event, during which oxygen was produced by living things at such a rapid rate it far exceeded the amount of oxygen that was captured from the atmosphere in Earth's geological/chemical reactions. The result of this shift (to the third atmosphere) has been the presence of free oxygen in our atmosphere ever since, providing the energetics to support the rapid evolution of life on Earth. And, of course, one of these life forms has become so dominant we've developed a way of life that is now fundamentally changing the atmosphere of our planet for a fourth time –but not for good. Now with this history of our planet's atmospheric evolution in mind, let's return to the earliest days of the Solar System and explore the very different path taken by Saturn.

Top: Stromatolites from the Cretaceous Period are the earliest fossil evidence of life on Earth and the first known organisms to photosynthesise and produce free oxygen.

Bottom: Before first life appeared on Earth, the planet had significant periods of volcanic activity.

Using a combination of geology, climatology and paleontology, scientists have been able to piece together an immensely detailed picture of the Earth's atmospheric evolution.

UNDER PRESSURE

Around four and a half billion years ago, the young Saturn had grown to become one of the largest rocky bodies in the history of the Solar System. At least ten times the mass of the Earth, it found itself far beyond the warmth of the Sun, locked in its icy orbit, a frozen rocky world on a scale unlike anything we have ever seen. But time would not stand still for this world; having grown as large as it could from rock and ice alone, its barren surface would not stay lit by the dim light of the Sun for long. As it continued to grow, Saturn would become a radically different kind of planet, defined not by its rocky surface but by an atmosphere that would bury it beneath vast volumes of gas, an atmosphere that would not just enwrap the planet but engulf it.

Just like the Earth in its earliest days, as Saturn continued to grow it drew not just solid material but also gas towards it from the vast clouds swirling around the embryonic Solar System. Huge amounts of the hydrogen and helium left over from the formation of the Sun clung to the vast planet, creating a first atmosphere that may well have been very similar to the earliest one on Earth.

'The actual formation mechanism is still up for debate, but once you get instability happening where you have enough material in the core, 10 to 15 Earth masses, it caused Saturn to grow to many hundreds of Earth masses.'
Kevin Baines, Cassini mission

ANATOMY OF SATURN

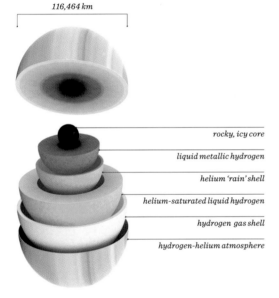

116,464 km

rocky, icy core

liquid metallic hydrogen

helium 'rain' shell

helium-saturated liquid hydrogen

hydrogen gas shell

hydrogen-helium atmosphere

But Saturn's early atmosphere would follow a very different course, as it was driven by two key differences between the young worlds. First, Saturn had grown far, far bigger than Earth, so by gravitational attraction alone, the volume of gas it could attract towards itself was far greater than anything the diminutive Earth was able to. But it wasn't just size that drove Saturn's atmosphere to grow so rapidly. Located this far from the Sun, the freezing temperatures meant that even the most volatile of gases could cling on to the planet without being driven off by the heat of the Sun. Hydrogen, helium, ammonia, methane and many other gases that were too light for the smaller worlds of the inner Solar System to hold on to were able to accumulate out here, with Saturn's vast mass and the freezing temperatures being enough to draw them in. Trillions upon trillions of tonnes of gas began to envelop the planet, reaching depths first of hundreds of kilometres and then thousands, and as this new atmosphere grew, it transformed the surfaces below.

On Earth, under our canopy of air, it's easy to forget that all of those molecules sitting above your head right now have a weight. If you draw a 1-centimetre square on a piece of paper (at sea level) and imagine a column of air stretching all the way up from the square to the top of the atmosphere, it will have a mass of on average just over 1 kilogram, exerting a downwards force of 10 Newtons onto the paper – a pressure that we call 100 kpascals, or 1 bar. That might not sound like much, but all of that weight reveals itself in the subtle and not-so-subtle effects of atmospheric pressure that we can see all around us. For example, the simplest of classroom experiments can reveal the power of the atmosphere by the process of just removing the air from a plastic bottle – the bottle collapses not because of the vacuum that it creates but because the bottle cannot withstand the pressure of the atmosphere that is pushing down on it.

While 100 kilometres of atmospheric pressure can quickly crush a bottle here on Earth, a much bigger atmosphere can exert pressures that not only are able to shape things on the surface but also have the power to transform an entire planet. And that's exactly what we think happened to Saturn, 4.5 billion years ago. As its atmosphere deepened, the thousands upon thousands of kilometres of gas weighing down on its surface began to generate incredible pressure, heating the rock and ice so much they began to glow. It's thought perhaps that for a brief moment in time the heat generated on the surface of Saturn by these extraordinary pressures meant that the glow of the planet even outshone the Sun. But this was just the beginning: as Saturn matured, the pressure at its core grew to 10 million times the level we experience on Earth. At that level of pressure, matter behaves in extremely strange ways, and the very idea of a planetary surface becomes meaningless.

As Saturn took on its new form, the enormous pressures generated crushed and melted its solid core, destroying the rocky world it had been and replacing it with a new world, utterly alien in its characteristics. Saturn had transformed from a world of rock and ice into a wholly different class of planet, a gas giant, a ball of gas so big it could contain 800 Earth-sized worlds inside it, making it the second-largest planet in the Solar System after Jupiter.

Opposite: Saturn's atmosphere is mostly hydrogen and helium; bands of clouds move across its surface at different speeds and directions, depending on their latitudes.

THE CASSINI MISSION

At 7.55.46 am Eastern Daylight Time, on 15 September 2017, NASA's Deep Space Network antenna complex in Canberra, Australia, received notification that contact had been lost with one of our greatest space explorers. After 13 years of extraordinary encounters, the Cassini spacecraft had made its last fateful manoeuvre, plunging headfirst into the atmosphere of Saturn. Forty-five seconds later the craft would be gone, crushed by the atmosphere of the planet it had spent over a decade so carefully exploring. Its last image, a monochrome view of the planet's night side lit by the reflected light of the rings, was a picture of its own grave, revealing the location towards which the spacecraft would plummet, in its final act of self-destruction as it entered the planet's atmosphere just a few hours later.

> 'Cassini was the most sophisticated mission we've ever launched to the outer solar system. It showed us a lot, but we need to go back, because we keep finding newer mysteries, even more exciting things about the moons and rings of Saturn.'
> *Kevin Baines, Cassini mission*

This was the end of a 20-year mission that had exceeded all expectations. The fourth spacecraft to enter the Saturnian system, following after Pioneer and the Voyagers, Cassini was the first to live there, ultimately spending 13 years and 76 days in orbit around Saturn. During this time it explored the planet, its rings and its moons and relayed images and information in a level of detail that would change our understanding not just of the Saturnian system but of the whole Solar System and our place within it.

T-minus 15 seconds
T-minus 10
9,8,7,6,5,4,3,2,1
And lift off of the Cassini Spacecraft
on a billion-mile trek to Saturn

It all began on 15 October 1997 at Cape Canaveral, with the launch of the spacecraft aboard a mighty Titan IV rocket. To get to this moment had taken the best part of 15 years, the combined efforts of 28 countries and a budget of $3.26 billion, but even then it would be another seven years before Cassini could truly begin its mission objectives. Getting this 5.5-tonne machine across 1.6 billion kilometres of space would require a complex trajectory that would take it on two flybys of Venus and out to the asteroid belt, before returning to the Earth two years after take-off for a final slingshot from its home planet that would send the craft hurtling at speeds of 98,346 mph outwards past Jupiter and on towards its final destination. This complex and high-risk trajectory was designed to exploit the orbital momentum of Venus, Earth and Jupiter, to enable the spacecraft to accelerate, which helped keep the amount of fuel and hence the launch weight to a minimum while making sure the craft could pick up enough speed to complete its epic journey. However, the trade-off of using this VVEJGA (Venus–Venus–Earth–Jupiter Gravity Assist) trajectory was a nervous six-year, 261-day wait for the team back on Earth as the craft made its convoluted voyage across space.

Primary voyage complete, Cassini now faced the life-or-death challenge of slowing itself down with enough precision to enter orbit around Saturn. The tiniest miscalculation would have seen billions of dollars flung into the outer reaches of the Solar System and decades of work lost forever.

'The Doppler has flattened out … we have arrived.'

On 30 June 2004 the spacecraft began its complex Saturn Orbital Insertion (SOI) manoeuvre by flying through the gap between the F and G rings. This manoeuvre alone was not without serious risk from the potentially devastating impact of particles in the rings colliding with the craft, so Cassini had to carefully orient itself to protect its instrumentation.

Once through the ring plane, the spacecraft had to rotate itself once again to point its engine directly along its flight path, so that when the engine fired it would decelerate the craft by a precise velocity, 622 m/s, allowing it to fall into the gravitational hand of Saturn. Decelerate too much and the craft would fall into Saturn; too little and it would fly off into outer space.

Left: Cassini's swansong. The spacecraft burned up in Saturn's atmosphere in a blaze of glory at the completion of a very successful mission.

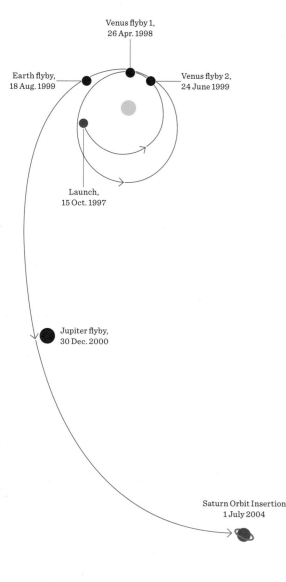

Venus flyby 1,
26 Apr. 1998

Earth flyby,
18 Aug. 1999

Venus flyby 2,
24 June 1999

Launch,
15 Oct. 1997

Jupiter flyby,
30 Dec. 2000

Saturn Orbit Insertion
1 July 2004

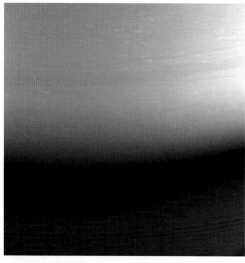

Top: Cassini's complicated flight path to Saturn from launch, which made use of gravity-assist flybys from surrounding planets.

Bottom: The last image taken by Cassini. It looks towards the planet's night side, which is lit by reflected light from the rings.

At 8:54 pm PDT, Cassini was finally captured by the gravity of this gas giant, becoming the first spacecraft to ever orbit Saturn. A new age of exploration had begun. And over the next 13 years Cassini would provide an extraordinary return on investment. The Voyagers and Pioneer had given us our first rough glimpses of Saturn's moons, but during its 13 years in orbit around the planet, Cassini did so much more. Not only did it discover a whole host of new moons, it also helped solve countless mysteries around moons we already knew existed, such as Titan and Enceladus, and revealed just how varied the characteristics of the 60-plus moons of Saturn really are. Cassini also allowed us to study for the very first time the interactions between Saturn's menagerie of moons and the ring system within which they sit, as well as the structure and history of the ring system itself. We'll come back to much of this exploration and revelation later in this chapter, but first we are going to take a look at what Cassini revealed about the structure of the gas giant itself.

One of Cassini's key mission objectives was to explore the characteristics and activity of the vast atmosphere that makes up most of Saturn. Voyager 2 had revealed the presence of a heat source buried deep beneath the planet's cloud tops that seemed to be producing vast amounts more energy – up to 87 per cent more, in fact – than the planet absorbs from sunlight, but what was the source of this heat and how did Saturn's immense atmosphere organise itself into the vast and intricate weather systems like the huge hexagon? This remained a question that Cassini was charged with solving.

Atmospheres can be some of the most spectacular and complex environments found on any planet. From the crushing intensity of Venus's choking atmosphere, with its raging winds and sulphuric acid rain, to the thin blue line of Earth where even on our somewhat milder planet we see endless storm systems. In the inner Solar System, the atmospheres of the rocky planets and the weather systems they contain are all powered by a single source – the relentless heat of the Sun.

All the weather we see on the rocky planets, whether it's a dust storm on Mars or a hurricane on Earth, are driven by the same simple process. It begins with the Sun beating down onto the planet's surface, heating the ground, which in turn radiates back, heating up the air closest to the ground. That air expands, which means it becomes less dense and so the hot air closest to the ground rises.

The result of all this rising air is that you get thermals – energetic movement of air from the ground into the upper atmosphere. Over the oceans exactly the same process occurs, with the heat of the Sun evaporating vast amounts of water and lifting it up into the atmosphere.

So in the very simplest terms, all of the extraordinary weather we see on Earth, or any of the other terrestrial planets, is driven by the Sun heating the land and sea and causing the air to move around. Travel further out into the Solar System, however, and the character of any atmosphere radically changes.

The storms on Saturn are among the most violent found anywhere in the Solar System. Generated by rich, complex and powerful weather systems, we've been witnessing these giant storms from afar for almost 150 years.

Asaph Hall, the great American astronomer and discoverer of the two moons of Mars, was the first to observe one of these storm systems back in 1876. Looking through the largest refracting telescope of the time (the United States Naval

Maybe a trickle of telemetry left ...
We've just heard the signal from
the spacecraft is gone ... And within
the next 45 seconds so will be
the spacecraft. I hope you are all
as deeply proud of this amazing
accomplishment. Congratulations to
you all. This has been an incredible
mission, an incredible spacecraft,
and you are all an incredible team.
I'm going to call this the end of
mission. Project manager off the net.'
Earl Maize, Cassini mission

Bottom left: After 15 years
of work and research, Cassini
receives its last checks before
launch.

Bottom right: Cassini
spacecraft about to dive
between Saturn and its
innermost rings as it nears
the end of its mission.

Top right: Cassini-Huygens
mission has liftoff, with
the spacecraft launched
on 15 October 1997 at Cape
Canaveral.

Top: An illustration of the storm on Saturn, January 2011 to March 2012, shows hot air hotspots in the planet's northern hemisphere.

Bottom: False-colour images of the largest and longest-lasting storm witnessed by Cassini, charted over the course of one Saturnian day.

'Saturn does explode into roiling storms every couple of decades; we were lucky enough to be there when one of the major storms ever seen on Saturn was spotted, which brought stuff up over a hundred miles high.'
Kevin Baines, Cassini mission

Observatory 66 cm telescope located in Washington DC) he was able to make out a great white spot sitting in the northern hemisphere of the planet. At the time, Hall wouldn't have known much at all about the phenomena he was looking at; it was intriguing simply because it enabled him to make the first estimate of the rotational period of the planet. Over the following decades, however, this great white spot was observed again and again, roughly falling into an appearance every 20 to 30 years before slowly fading away. Fortunately for us, the long gaps between each appearance were cut short in December 2010, when a great white spot appeared just four years after the previous one. This time, with the Cassini probe in orbit, we had the opportunity to track the evolution of one of these storms close up, from its birth in early December 2010 to its disappearance eight months later. This vast storm system, resembling in many of its characteristics a simple thunderstorm here on Earth, emerged as a single white spot 1,300 kilometres from top to bottom and 2,500 kilometres from side to side. Over the ensuing months Cassini was able to follow the storm as it stretched out around the planet. Just a few weeks after its formation, the tail of the storm had extended across 100,000 kilometres, and within six months this great storm had stretched so far it encircled the whole planet. Cassini was also able to image directly for the first time lightning occurring in the middle of this giant storm in broad daylight. The fact that the lightning could be detected on the day side of the planet surprised the Cassini team and suggested that the lightning must be particularly intense to make it visible in such bright light.

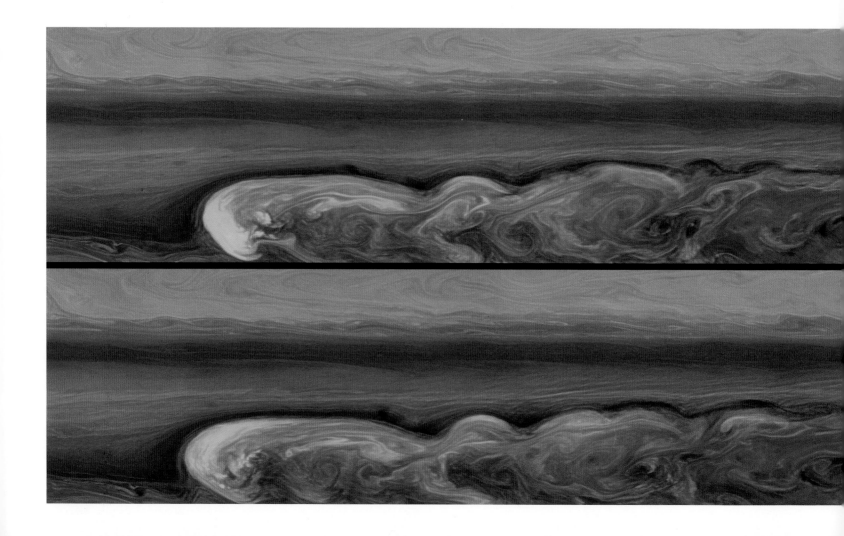

Left: This series of six images from Cassini shows the development of the storm, from small white cloud to indistinct formation in the bottom right image.

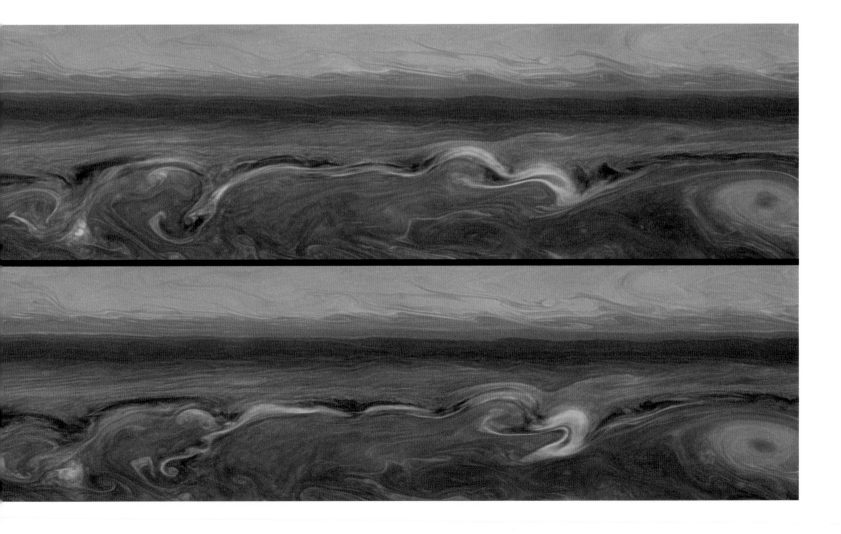

'Cassini could even listen to the crackling of the lightning strikes that were taking place within this powerful convective storm, some 10,000 times more powerful than anything we get here on Earth.'
Leigh Fletcher, Cassini mission

'The lightning flashes are comparable to the strongest lightning flashes that we see on Earth. That was a joy, just to finally witness lightning on Saturn, and we could tell that it was coming from the cloud layer, the deepest cloud layer, which is about 60 miles down, and is the water cloud; the water ice cloud. And that makes sense because lightning is produced on Earth ... because of updrafts and particles being elevated through an atmosphere, rising through an atmosphere because they're convecting.

We had a whole variety of scientific goals at Saturn, and one of the main ones was to understand the meteorology of the Saturn atmosphere and what energises the winds that we see on it, and so on. And we have affirmed now the belief that the atmospheric systems on Saturn are actually powered by energy from below; from an internal source on Saturn. They're not powered by sunlight, like we have on Earth.'
Carolyn Porco, Cassini mission

Cassini was also able to turn its cameras onto the giant hexagon structure that Voyager 2 had first photographed over 30 years before. This time with its high-definition, narrow-angle cameras trained on the enormous feature around the north pole of the planet, Cassini was able to capture the giant hurricane-like storm in extraordinary detail. This vast vortex was measured as 2,000 kilometres across at just the eye of the storm, and with cloud speeds travelling at over 150 m/s it is rotating far faster than any hurricane we have ever seen on Earth.

Exactly what causes this or any of the giant storm systems we have witnessed on Saturn is still not fully understood, but one thing we absolutely do know is that this far out from the Sun these weather systems cannot be driven primarily by the Sun's heat. With Saturn receiving 100 times less of the Sun's energy than Earth, something else must be driving its weather, something that creates the same great upswellings of heat we see on Earth but that on Saturn is entirely generated from the interior of the planet.

As yet we have not been able to peer deep enough into Saturn's interior to see directly what happens inside this strange world, but in the last couple of years we've come closer than ever. As we have accumulated evidence from Pioneer, Voyager and of course Cassini, and combined it with our increasingly sensitive ground-based observation, we have been able to start piecing together a detailed journey into the heart of the planet. By just touching the cloud tops of Saturn, alive with great storm systems and flickering with immense bolts of lightning, they have given us the first hints of the truly strange world that must lie beneath and the mysterious energy source that brings this planet to life.

Top: A storm circling around the north pole of Saturn. The eye of the storm is about 50 times larger than the average hurricane eye on Earth.

Right: White arrows in this mosaic of a storm on Saturn reveal lightning strikes amongst the clouds. Lightning is present on the left and absent on the right.

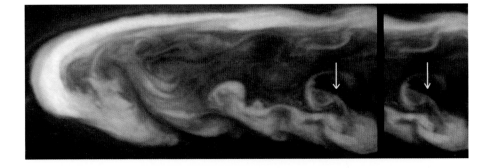

TO THE VERY LAST

After 20 years in space, this battered explorer was no longer able to maintain a stable flight path through the increasingly dense atmosphere. Beginning to tumble and roll, the last few bytes of data left its antenna streaming the 90-minute journey back to Earth.

Our journey deep into the interior of Saturn begins not with clouds but with rain. At the very outer edges of its atmosphere, in the space between the closest ring (the D-ring) and the planet itself, we have discovered the phenomenon of 'ring rain'– an endless downpouring of material falling from the rings into the upper atmosphere of the planet below. We know this because as Cassini took its final, fatal plunge on 15 September 2017 it flew through this unchartered no man's land and was able to make the first direct measurements of the rain falling from the rings. Using the Ion and Neutron mass spectrometer, it measured the rate at which the rain was falling from the rings – a staggering 10,000 kilograms of material every second. It was also able to reveal that the ring rain is comprised of a mix of hydrogen, water ice and a surprising quantity of complex organic compounds like butane and propane, all showering down onto the planet and changing the chemistry of its upper atmosphere in ways we are only just beginning to understand.

Cassini then began its final historic dive into the upper atmosphere of Saturn, struggling to keep its antennae pointing towards the Earth as it was buffeted by the increasing turbulence, while continuing to send a stream of precious data back to us right until the very end. Travelling at around 123,000 km/h, Cassini entered Saturn's upper atmosphere at 3:30:50 PDT, 1,900 kilometres above the planet's cloud tops. This was the first time scientists had ever been able to sample the atmosphere directly, our distant human envoy touching the gas giant for the very first time. Over the next few seconds Cassini increased the use of its altitude control thrusters to steady its rapid descent into the thin outer reaches of Saturn's atmosphere and keep the antennae pointing towards Earth for a final few seconds of contact. But this was a losing battle; over the next minute and a half, the growing density of the upper atmosphere rapidly began to increase the friction on the forward-facing surfaces of the spacecraft, raising the temperature at high speed and creating an unfolding thermodynamic situation that Cassini had not been designed to withstand.

At precisely 3:30:00 PDT, 1,500 kilometres above the cloud tops, Cassini finally lost its battle for control. After 20 years in space – 13 of which were spent exploring Saturn – and having travelled a distance of 7.9 billion kilometres, this battered explorer was no longer able to maintain a stable flight path through the increasingly dense atmosphere. Beginning to tumble and roll, the last few bytes of data left its antennae streaming the 90-minute journey back to Earth, and the mission was long over before that final contact reached our planet.

Precisely what happened next to the little spacecraft as it fell into the grip of this giant planet remains no more than well-informed conjecture. Using computer models to predict its final seconds, the Cassini team were able to watch only in their mind's eye as the craft tumbled chaotically through the atmosphere. Travelling at 144,200 kilometres per hour, just 1,100 kilometres above the cloud tops and with temperatures rapidly increasing, every component of this unprotected machine began to disintegrate. With all computer systems failing and the spacecraft blind and deaf to its surroundings, it began to break up, its gold insulation blanket rupturing first, before the external carbon fibre structures of the spacecraft itself began to fracture and fragment. Then with the rings gleaming in the sky above it and the yellow cloud tops beckoning below, what was left of Cassini streaked across

Above: Scientists at NASA's Jet Propulsion Laboratory bid a tearful goodbye to Cassini.

Opposite: For its grand finale, Cassini made 22 orbits that swooped between the rings and the planet. It ended its mission on 15 September 2017 with a final plunge into Saturn's atmosphere.

<u>Jonathan Lunine on the importance of the</u>
<u>Cassini mission</u>
'Cassini will go down in the history of planetary
exploration as one of the most remarkable missions,
particularly in the sense that it not only made
discoveries, but was able to follow up on its own
discoveries with genuinely new observations.

Cassini discovered the plume of Enceladus,
it detected the presence of the ocean, and it was able
to follow up and actually fly through the plume and
then with a whole different set of instruments measure
its composition. That's something that normally we
expect would happen on the next mission.

Cassini discovered the lakes and seas of the northern,
high northern latitudes and the southern latitudes
of Titan and was able to follow up by using the radar in
a different way, so we could actually probe the depth
and composition of those seas. Again, you would expect
that would happen on the next mission. But Cassini was
able to all these things as if it was a series of missions,
not just one mission, and that was what was remarkable
about it.'

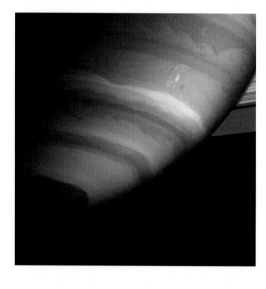

Above and opposite: The
mission to the never-before-
explored region between the
planet and its rings answered
many of NASA's questions,
but have posed many more.

the Saturnian sky, blazing with temperatures higher than the surface of the Sun. The gas tanks of propellant that helped steer the craft on its epic voyage blasted apart the remaining structures into a million different pieces, which in a final spectacular moment lit up a little corner of the majestic planet with the burning debris of its greatest explorer.

Cassini was dead, but its final journey was far from over. Over the following minutes the debris scattered and fell through the cloud tops, deep into the planet's interior, until finally the atoms of this beloved machine, built by human hands millions of kilometres away on a faraway world, came to rest in the heart of the planet. Cassini and Saturn, inseparable for evermore.

Where the debris of Cassini finally ended up we can only imagine, but the data from this extraordinary mission that was collected right up to its very end has helped us dive deeper into the detail of Saturn's interior than ever before. As the wreckage scattered and fell from the upper atmosphere it would have fallen through the very first cloud tops of the planet. Here, with temperatures ranging from minus 170 to minus 110 degrees Celsius, the clouds are made not of water ice, but frozen ammonia. It's these ammonia ice crystals that give Saturn its characteristic pale yellow colour, a colour that we can see shining back at us with even the most modest of Earth-bound telescopes. Beyond this upper layer of cloud, as the pressures begin to increase and the temperature rises, huge clouds of water ice form a thick layer, the water molecules perhaps long-lost remnants of the building blocks of the icy world that Saturn once was. It's here, amongst the vast cloud systems, that lightning more powerful than anything we see on Earth illuminates the sky within storm systems that rage for months on end. Cassini gave us the first still and moving images of these lightning storms and enabled us to predict the next great transformation that will happen within the planet's interior.

'Every couple of decades, Saturn surprises us and breaks into a fit, and actually erupts with really major storms. These are global storms … And they last for about six months to a year. The one that we saw with Cassini in 2010 was so powerful that our calculations show it was the most powerful storm we've ever seen anywhere in the Solar System, even Jupiter, in terms of its central conductive storm system.

We had an amazing chance to see that during the Cassini mission. And then when we looked at those clouds that it was forming, we found amazing things in the cloud, in the storm clouds of Saturn … We found that it had water ice and this really blew our minds because water should be sequestered deep down in the atmosphere. You should not find it at the top of the atmosphere. It only gets there when you have huge convective updrafts, and we're talking … updrafts that bring gas up from 100 miles down. And on the Earth, we only see gas going up and convection going up only about 10 miles. Here you have storms 10 times bigger. Very powerful storms, lifting things up from 100 miles down, and bringing up the water from below, which was probably actually a mixture of gas and water. But then it brought it up and it turned into ice, and the component of water ice in it really shocked us.'
Kevin Baines, Cassini mission

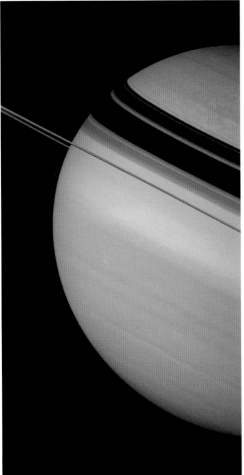

We now think that around 100 kilometres below the visible atmosphere (the same distance from the surface of Earth to space), with temperatures inside Saturn's lightning reaching almost 30,000 degrees Celsius, this lightning transforms methane gas in the atmosphere into huge clouds of carbon soot. Carried higher into the atmosphere on great convection currents, we think that it's these specks of carbon-producing lightning that can explain the mysterious dark clouds that Cassini was able to observe and image over many years. This particulate matter rains down from the upper cloud layers until, at a depth of 1,500 kilometres, we think the enormous pressures begin compressing this soot into graphite, which then falls further into the depths of the planet. We then think that at around 6,000 kilometres down, this graphite undergoes a truly extraordinary transformation. With pressure and temperatures rapidly increasing, the graphite is compressed into solid diamond that rains down for at least 250,000 kilometres. Estimates suggest at any one time 10 million tonnes of diamond 'hail' is raining down within the planet, from specks of diamond dust just a millimetre across to chunks as big as 10 centimetres. Perhaps even for a brief moment in time, some of the charred carbonised remains of the Cassini probe made it down into these depths and transformed into diamonds as well, falling further into the planet it had helped us understand. But eventually even these diamonds will ultimately succumb to the enormous pressures of Saturn's interior. Around 30,000 kilometres into the atmosphere, where temperatures reach 8,000 degrees Celsius, the diamond melts and forms liquid diamond raindrops that continue to descend into the abyss.

It's here, around 40,000 kilometres down into Saturn's interior, that we believe the source of Saturn's energy, the energy that drives all the storms and weather we see playing out on its surface, is finally revealed. Here the pressures are so intense that the hydrogen that makes up over 96 per cent of Saturn's mass is compressed into a vast ocean of liquid hydrogen that is saturated with helium gas. Under these enormous temperatures and pressures the helium precipitates out and falls like rain through the ocean, depleting the levels of helium in Saturn's outer layers (which explains the lower than expected amount of helium first measured by Voyager 1 in 1980). It's a process that through the simple action of friction as the molten helium rain falls creates an incredible amount of heat and we think this is the engine, the energy source that helps power Saturn's ferocious weather and drives its entire atmosphere for tens of thousands of miles in every direction.

Finally we reach the interior of Saturn, a layer where the liquid hydrogen is so compressed it starts to behave like a liquid metal and the helium itself may accumulate into a metallic shell surrounding the core. Beyond here we do not know if a rocky core remains, as a remnant of Saturn's long-lost past, or whether under these enormous pressures the possibility of solid rock is no longer viable. Whatever lies at the centre of Saturn, we estimate that this core region has a mass of around 20 times that of the Earth, and a diameter of around 25,000 kilometres. And here, perhaps, a few atoms of Cassini might have come to rest in the ancient heart of the planet.

SATURN'S ICY RINGS

Since its earliest childhood years, Saturn has lived a life of incredible drama. The evolution from rocky beginning to the extraordinary gas giant we see today took just a few hundred million years of rapid transformation. And yet throughout almost all of this time, one thing has been missing. We now think Saturn was born naked and has lived out the vast majority of its existence without the structure that makes it more iconic than anything else.

Saturn is a world defined by its rings. Ask a child to draw a planet and more often than not they'll draw something that looks remarkably like Saturn. As we discovered earlier in the chapter, the particles in the rings of Saturn are almost impossibly delicate and intricate; most of them are no bigger than snowflakes but some are as large as houses, orbiting the planet at 1,800 km/h in one of the most beautiful structures we have ever laid eyes on. One of the reasons why the rings have captivated us for so long is that, despite their vast distance from Earth, if you look through a simple telescope they shine back at us, bright against the night sky, reflecting the Sun's light as powerfully as the planet itself.

'Saturn's rings are truly extraordinary. They're actually made up of separate ice particles, and the thought that millions of individual particles, each on its own path, can combine to create such intricate, beautiful waves and structures is astonishing. Some are only the size of tiny marbles while others are the size of mountains.'
Linda Spilker, Cassini mission

For years, the age and origin of the rings has been argued over, with conflicting evidence supporting two very different hypotheses. On one side the argument has been made that a ring system as dense as Saturn's must be an ancient structure shaped from the remnants of the planet's formation. On the other side of the fence the evidence has been used to suggest that the rings must be youthful, a relatively recent addition to the ancient planet.

The brightness of the rings is not only beautiful but also rather surprising. The Solar System is not a very clean place, in fact it's full of dust and dirt, and so we would expect any ice crystals that hang around for any length of time to inevitably get dirty. We also know that if ice gets dirty it doesn't reflect sunlight anywhere near as brightly, so in the case of the rings we would have expected if they had been present over the lifetime of the planet all of those particles and blocks of ice would have become increasingly less reflective. If they had formed anywhere near the birth of Saturn they would almost certainly be invisible to us now, or at least incredibly faint. And yet they shine brightly, as if they were made just days ago.

Above: Saturn's moon Mimas
set against the planet's rings
A, B, C and F.

Variations of this debate have been rumbling since the eighteenth century, but that all changed with Cassini. During its 13-year tenure around the planet the probe was able to not only see Saturn's rings like never before but also provide us with direct evidence about their density and composition. And this information is helping scientists to solve the puzzle of the rings once and for all.

For the very first time, Cassini was able to take physical samples directly from the rings and surrounding dust using the Cosmic Dust Analyser (CDA) mounted on its exterior. This highly sophisticated piece of technology is basically a multi-million-dollar bucket, but a bucket that can detect particles one-millionth of a millimetre wide (that's smaller than a virus) and determine the particles' size, speed and direction of travel as well as their chemical composition. By measuring the dirt and dust from the Saturnian system falling onto the rings, Cassini has allowed scientists to rewind the clock on the rings and provide direct and compelling evidence that the ice particles are nowhere near as dusty as they would be if they were an ancient part of the planetary system. This confirms the long-held belief that the rings are so bright because they are young, and in fact the evidence gained from Cassini's CDA suggests they are far younger than we ever imagined. Latest estimates suggest they are somewhere between 10 to 100 million years old – that's nearly 4.5 billion years younger than Saturn itself. So where did they come from? Once again, Cassini provided a tantalising clue.

In its final few days, Cassini embarked on a series of almost reckless orbital trajectories that were so risky to the spacecraft's safety they would never have dared to be attempted earlier in its mission. Taking the probe closer to the planet than ever before, the team flew Cassini repeatedly through the gaps between the upper atmosphere of the planet and the inner rings. Throughout September 2017, as it performed 22 of these perilous dives, Cassini was able to take measurements of the gravitational tug of both the planet and the rings. This data was then used to calculate the mass and density of the most substantial ring in the Saturnian system. The B-ring makes up 80 per cent of the mass of all the rings and yet Cassini's measurements suggested the total mass of the ring was surprisingly low. This result, although not conclusive, strongly points to the fact that the rings' origins lie not in the ancient past during the formation of the planet itself, but to a far more recent and dramatic event involving one of the worlds trapped in orbit around it. The rings, it seems, were born of a moon.

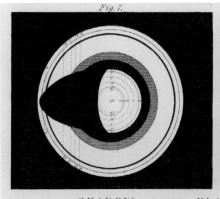
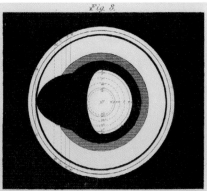

SHADOW OF THE PLANET ON THE RINGS
AT DIFFERENT SEASONS OF THE SATURNIAN YEAR.

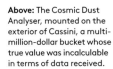

Above: The Cosmic Dust Analyser, mounted on the exterior of Cassini, a multi-million-dollar bucket whose true value was incalculable in terms of data received.

Right: Engraving from 1865 featuring the shadows of Saturn's rings, revealing how far back in time the fascination with this mysterious planet stretches.

SATURN'S ICE RINGS

Ice is like the outer Solar System's rock. The moons of the outer planets are largely made of ice. We don't have as much ice here in the inner Solar System where it was warm and it vaporised and it didn't collect, but out there it was cold enough when the planets and their moons formed that lots of ice condensed out of the cloud and gas and dust from which the planets were being born. So there's plentiful ice on the surfaces and in the interiors of the moons of Saturn and the other outer planets of the outer Solar System. The ice is a fundamental building block of these outer worlds. Metal, rock, ice. And ice is what we see on the surface, it's the least dense stuff. A lot of these worlds were warm enough that the rock and the metal went to the interior and the ice is on the outside, and that's what we see and that's the

geology we study. We study the geology of frozen, very cold ice that behaves like rock because it's that cold and the ice is that hard out there in the outer Solar System.'
Bob Pappalardo, Cassini mission

'If I could hold a ring particle in my hand, from the data I have in the infrared, it looks like probably the outside is very fluffy and porous, maybe kind of like snow, and perhaps on the inside there's an icy core surrounded by this very fluffy exterior. And we now know that ring particles tend to stick together and make long chains of particles that stick together for a while and then break apart and float around.'
Linda Spilker, Cassini mission

Opposite: This density wave captured here is caused by orbital dissonance with the moon Janus.

Bottom: The close-up images from Cassini revealed vertical structures along the edge of Saturn's B-ring, which tower 2.5 km high.

Below: Cassini's images reveal elaborate and complicated patterns in Saturn's rings, created by their varying opacity and the shadows they cast.

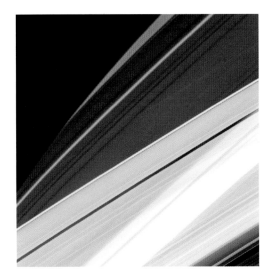

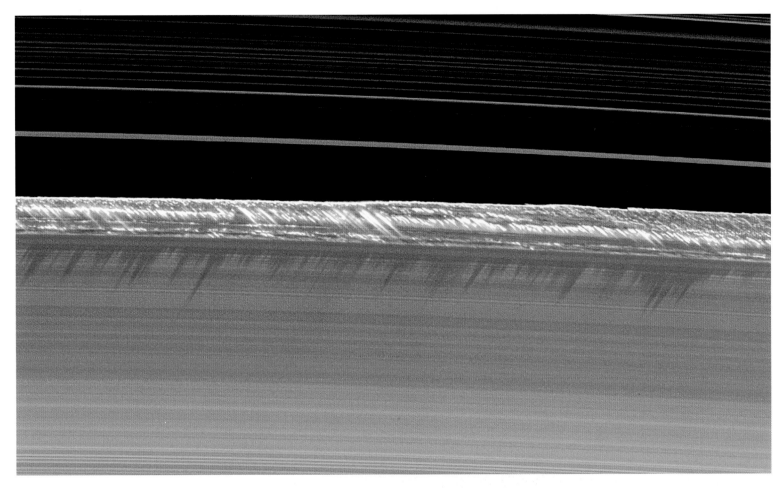

SATURN'S MOONS

'Saturn's moons are the archaeological debris from planet formation; their size, their orbits, even their composition help us to understand the different epochs of the evolution of the planet.'
Jonathan Lunine, astrophysicist

Saturn has 62 large moons and countless smaller ones that range in size from planetary-scale objects like Titan to small irregular lumps of rock like Aegaeon. And we're still hunting for new moons to add to the list. On its travels, Cassini spotted two objects within the F-ring that are potential candidates for 'moonhood', but as yet S/2004 S6 and S/2004 S3 remain unverified and unconfirmed.

Within this menagerie of worlds are an endless variety of clues written into the size, form and surface features of the moons that allow us to read the history of the Saturnian system, revealing the often violent and volatile events in its past.

Mimas is otherwise known as the Death Star moon, and was first spotted in 1789 by William Herschel, who is immortalised in Mimas's eponymous huge impact crater. Mimas is just 396 kilometres across and the Herschel crater covers 130 kilometres of that surface – a deep scar that tells us that once, long ago, there was a giant impact that almost destroyed the moon.

Iapetus, the third-largest of Saturn's moons at 1,500 kilometres across, is almost planetary-sized and has a distinctive two-tone colouration – half light and half dark. Iapetus was discovered by Giovanni Cassini himself in 1671 and has another distinctive feature that we have long puzzled over. Running around three-quarters of the moon's diameter is a massive equatorial ridge. It's one of the highest mountain ranges in the Solar System, 20 kilometres from the peaks to the ground below, and it's thought this may have formed because a ring that once circled the moon collapsed onto its surface.

Hyperion is the strange sponge-like moon that tumbles around in Saturn's orbit, which, in 1848, was the first non-spherical moon ever to be discovered. To this day it remains a mystery as to what caused its punctured surface and irregular shape, and it has been suggested that Hyperion is in fact a remnant of another moon that broke up long ago.

Pan, the small walnut-shaped moon that was discovered less than 30 years ago from the analysis of images taken by the Voyager 2 probe, is named after the Greek god of shepherds. It is so named because Pan is a shepherd moon, clearing the orbit of the Encke Gap, a 325-kilometre space within the A-ring. Its distinctive equatorial ridge is thought to be made of material that has been swept up and accumulated as it clears its path through the ring.

Cassini's endless stream of images and data until 2017 has enabled scientists to analyse each of the moons in ever more detail, and it's become increasingly apparent that they are intimately linked with the ring system they orbit. Many of the moons are made of exactly the same material as the rings themselves – Mimas, Iapetus and Hyperion are ice moons composed almost entirely of frozen water ice, with only the smallest amounts of rock in their makeup. From the evidence that we have been able to read on their surfaces it seems almost certain that through 4 billion years of violent interactions at least some of the moons of Saturn must have both come and gone over the planet's history. All of this begins to point us towards the idea that perhaps the most intriguing moon in Saturn's long history is one that is now entirely missing.

Opposite: Dione and ring shadows on Saturn. Dione is around 1,123 km in diameter and orbits some 377,000 km from the giant Saturn.

Left: Two images show two perspectives of Pan as Cassini passed by in its closest ever encounter with the little moon.

Top left: Cassini captured this detailed image of cratered Mimas in 2010, when the spacecraft came within 9,500 km of the moon.

Below: The clearest views ever taken of Saturn's largest moon, Titan, which is usually camera shy and shrouded in haze.

Around 100 million years ago, as dinosaurs roamed the Earth, we now believe a long-lost moon orbited close to the planet Saturn, a moon perhaps 400 kilometres across and formed almost entirely of ice. This was a doomed world, orbiting too close to resist the immense pull of Saturn's gravity, its fate sealed at the moment of its birth.

It was the French astronomer and mathematician Edouard Roche who first suggested the existence of this long-lost moon in 1848. He christened it Veritas, after the Roman goddess of truth, who according to legend is supposed to have hidden in the bottom of a holy well. Roche had postulated the existence of this moon while exploring a particular dynamic of celestial mechanics that describes the potentially violent gravitational relationship between two celestial bodies. He was able to demonstrate that when a moon moves within a certain distance of the planet it is orbiting, it can cross a threshold where the force of gravity is so strong the moon will be ripped apart.

This distance has become known as the Roche limit, and is dependent on many factors, including the mass, radius and physical characteristics of the moon and planet. Outside of the Roche limit a moon can exist in happy equilibrium with its planet, orbiting in the gravitational grip of the larger body while maintaining its physical integrity. But if that same moon's orbit degrades to the point where the Roche limit is crossed, the tidal forces exerted on the moon exceed the gravitational attraction that holds the moon together in a sphere, resulting in the moon's destruction.

This relationship applies to any moon/planet system, including our own. Here on Earth we see the relationship between our planet and Moon on a twice-daily basis as the tides ebb and flow across the planet. These tides are caused by the difference in the gravitational pull of the Moon from one side of the Earth to the other, and that difference – although subtle – is powerful enough to move entire oceans with the forces of tidal gravity.

Above: Saturn's sponge-like moon Hyperion, a world covered with strange craters.

Above: Saturn's third-largest moon, Iapetus, is known as the yin and yang, because of its dark and significantly lighter hemispheres.

'By understanding how other worlds work that are familiar yet different we better understand how our own Earth works. If you have an idea of how the surface of Earth is shaped, or how it evolved, you should be able to apply that to an icy world.'
Bob Pappalardo, Cassini mission

Just as the Moon creates tides on Earth, so the Earth also has a tidal effect on the Moon, and because the Earth is much greater in mass that effect is more powerful than you might think. The Earth's pull is in fact powerful enough to deform the Moon's surface. Even today it causes moon tides on the lunar surface, where it causes the rock to rise and fall, but the effect was even stronger 4 billion years ago when the Moon was nearly 17 times closer to the Earth. Back then, the pull from our planet caused a tide of solid rock to rise and fall by many metres, and if the Moon had ever come any nearer, it would have crossed the Roche limit and been ripped apart.

When Edouard Roche first formulated the equations calculating this line, he realised that the result of such destruction would not just be the obliteration of a moon but also the creation of millions upon millions of fragments, the debris of the moon orbiting around its destroyer. He also deduced that in time these fragments would form once again into the most delicate of structures – a system of rings – under the influence of the planet's gravity.

In one of the more gruesome tales from classical mythology, Saturn, the Roman god of time and harvest, actually ate five of his newborn babies to prevent them taking his power. Now Cassini has shown us that, at least metaphorically speaking, they weren't far wrong.

We now think that 10 to 100 million years ago, just beyond the outer reaches of Saturn's vast atmosphere, an icy moon orbited, gradually edging closer and closer to that tipping point, the Roche limit. As Saturn's immense gravitational force pulled at it, this great moon began to rupture, and in a moment undoubtedly worthy of even the greatest of mythological stories, up to 30 million trillion tonnes of ice, more than 30,000 times the weight of Mount Everest, broke apart in orbit and spread out around Saturn. Travelling at tremendous speed around the planet, it's likely that this debris quickly scattered to encircle the giant planet, and in the space of no more than a few days Saturn's iconic rings had been born.

Today this giant ring system has evolved into the intricate structures we can see in the night sky. Saturn's powerful gravity has helped keep its near-perfect elliptical shape, but an endless series of collisions have caused it to gradually flatten out. Now this debris forms a disc wider than Jupiter yet on average just 10 metres thick. With moon-sized chunks of ice orbiting through this structure, great voids have been cleared, turning one ring into many. In places, moons have pulled particles of ice upwards to create strange peaks over a kilometre high, which cast spectacular shadows across the rings.

Saturn, a once tiny world of rock and ice, a world that has undergone repeated transformations, has today become the Solar System's greatest jewel. And Cassini, our most intrepid of explorers, has forever deepened our understanding of the life story of this planet.

Yet perhaps Cassini's greatest gift is not in its revelations of the past but in the glimpse it has given us of the future, because just beyond Saturn's rings, Cassini would discover a hidden treasure, a world that may hold answers to some of our deepest questions about our place not just in the Solar System but in the Universe.

Above: Saturn, a ruthless god determined to rule over his domain, which perhaps has parallels with Saturn and its moons.

Opposite: Our balanced ebb and flow of the tides is down to the harmonious relationship between Earth and our Moon.

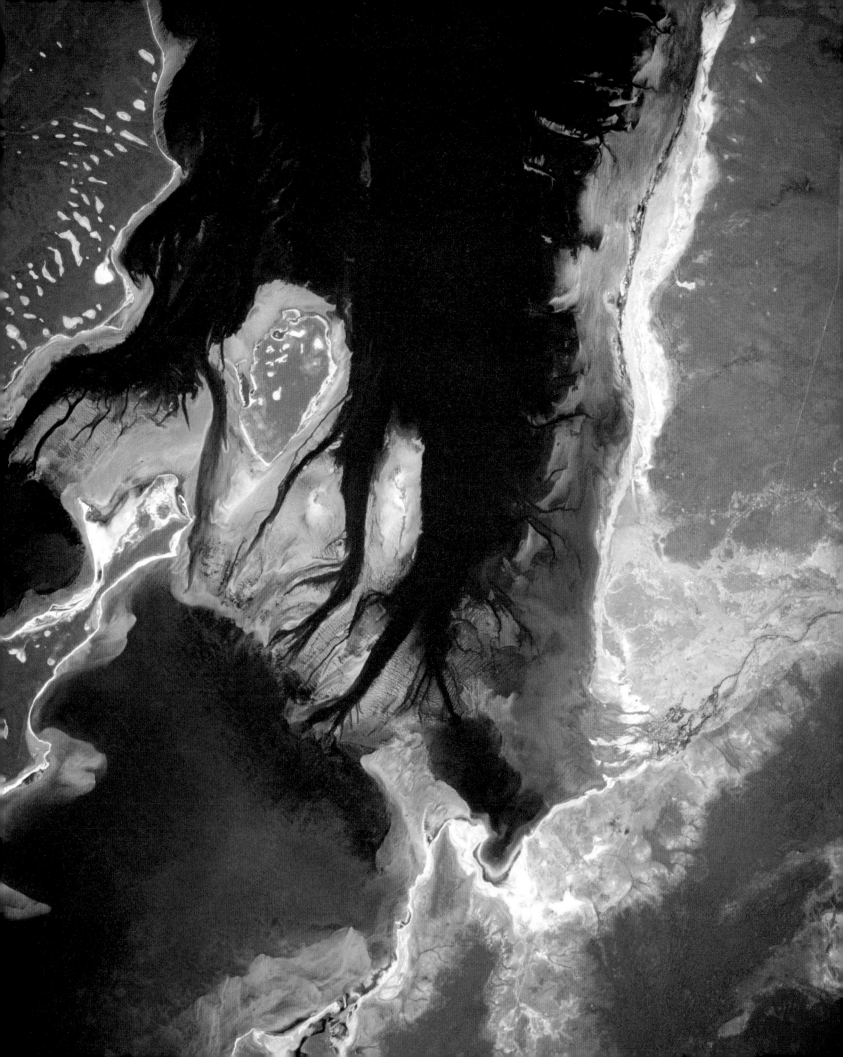

SATURN'S MOONS

I remember when Cassini launched, we knew Saturn had 18 moons. By the time we got there, I think we were into the 50s, and then Cassini added quite a few. I think Cassini added six. So all of the other ones, in fact the majority, the 60-plus moons that Saturn has, are actually these outer moons, which the giant planets, particularly Jupiter and Saturn, seem to have. These are almost certainly captured objects, captured Kuiper Belt objects, perhaps asteroids. The moons tend to be in virtually coplanar orbits, and the objects themselves are near-spherical. They tend to sort of occupy the middle ground, so not really close to the rings, but not too far out either.

There does seem to be a tendency, particularly in the Saturn system, and I think Uranus and Neptune as well, that the further out you go, the bigger the moons are.

Then once you get in towards the rings you find lots of small moons, which you would call, I'd say, the irregular satellites. But ring moons, which either have their own ring that is formed from impacts, and material that's spread around them or their orbit, or they're influencing a ring – they've created a gap for themselves, and they're influencing the material on either side.

And over and above that, there are these lovely little moons that are in the same orbit as some of the bigger moons. So, Tethys has an object 60 degrees ahead of its orbit, and another 60 degrees behind. And then Dione has an object 60 degrees ahead, and Cassini found an object that's 60 degrees behind. So we understand why these occur in some moons, and not in others.'
Carl Murray, Cassini mission

Below left: Three of Saturn's moons captured together: Titan, Mimas and Rhea, all seen as crescents.

Below right: Pale, icy Dione, set against the rings of its planet, Saturn.

Bottom: Prometheus casting a shadow on the F-ring of Saturn as the planet nears its August 2009 equinox.

Opposite: Daphnis, one of Saturn's ring-embedded moons, creates waves as it orbits within the Keeler Gap, disrupting the tiny particles of the A-ring with its gravity.

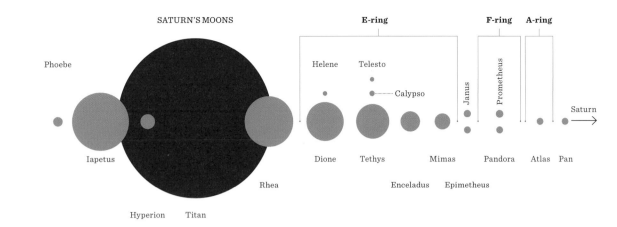

SATURN'S MOONS

E-ring **F-ring** **A-ring**

Phoebe

Iapetus

Hyperion Titan

Rhea

Helene Telesto

Calypso

Dione Tethys

Enceladus Epimetheus

Mimas

Janus

Prometheus

Pandora

Atlas Pan

Saturn →

VISITING TITAN

One of the unmistakable jewels of the Saturnian system is Titan. A giant, it is the second largest moon in the Solar System and this planet-sized satellite is larger than the smallest planet, Mercury, and 50 per cent bigger than our own solitary Moon. Discovered by Dutch astronomer Christiaan Huygens in 1655, it has held its secrets close for centuries, cloaked by thick clouds. The only moon in the Solar System with a dense, distinct atmosphere of its own, from afar Titan hinted at being far more like a planet than a moon. It would take a mission of breathtaking ambition to finally peer under her veil and see the wonders that lay beneath.

On Christmas Day 2004, after a seven-year journey across the Solar System, the Cassini probe released its sibling craft Huygens, setting it off on a 4-million-kilometre path to Titan. Simple in appearance, Huygens looked as if it had borrowed its design from the classic cartoon UFOs of the 1950s. But nothing was simple about Huygens; this was a highly advanced spacecraft about to begin one of the most complex manoeuvres ever attempted in the history of human space exploration.

As it made its three-week journey towards Titan, the team of ESA (European Space Agency) scientists, who had dedicated a lifetime's work to the project, watched on, hopeful but helpless. From the moment of release they had lost direct control of the probe and were unable to send it commands; instead, as planned in the mission protocol, Huygens was now entirely controlled by the on-board autonomous computer systems. All they could do was watch and wait to see if Huygens could fulfil their ambitions by becoming the first probe to land in the outer Solar System, on a moon other than our own, in the most distant landing of any spacecraft.

To succeed, Huygens would have to execute a perfectly controlled landing protocol designed to use Titan's atmosphere as an aerobrake, before employing a parachute system to deliver the scientific instrumentaion to the surface. And all after awakening itself from a 6.7-year intergalactic sleep. The probe had remained in an entirely dormant mode except for a six-monthly self-health check, and only at the point of separation from Cassini was the timer set to wake the probe up 15 minutes before entering the atmosphere of Titan. The proposed landing site for Huygens also held much uncertainty, chosen to be in the southern hemisphere of Titan (192.3 degrees west, 10.3 degrees south, to be precise). The area was surveyed by Cassini's cameras from an altitude of 12,000 kilometres and appeared to be near a site that had many of the characteristics of a shoreline. Landing into an extraterrestrial ocean had been part of Huygens's original design, so in theory the craft was expected to survive any splashdown into a liquid surface on Titan.

On 14 January 2005, the 318-kilogram probe entered Titan's atmosphere, 1,270 kilometres above the surface, and began its descent protocol. On-board accelerometers monitored the craft's deceleration, waiting for the precise moment to trigger the explosive bolts that would blow off the front heat shield and back cover and release the pilot parachute, which would then pull out the main parachute, slowing the 2.7-metre-wide craft from its nearly 2,000 km/h descent and allowing it to drift safely to the surface. All of this would need to play out perfectly while, more than a billion kilometres away, a group of nervous humans sat in near-silence in mission control waiting for the first telemetry to confirm that Huygens was safely on the ground. Every second was utterly critical because Huygens had no more than three hours of battery life and with an expected 150-minute descent that meant not

ORBITING TITAN

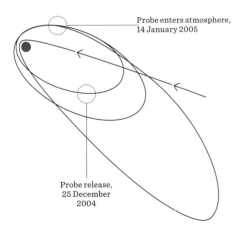

Probe enters atmosphere,
14 January 2005

Probe release,
25 December
2004

Opposite: Titan's dust storms as imagined in an artist's impression. These are believed to whip up during powerful methane storms in the moon's complex atmosphere.

Above: Reaching Titan was no easy feat; to get Cassini-Huygens to land safely on the moon required complex manoeuvres that took advantage of four planetary flybys to slingshot the spacecraft into position.

'A lot of the imagery we got on the way down once that haze cleared, opened up our eyes to this magnificent and amazing very Earth-like world. You have drainage channels, you have highlands, you have lowlands, you have seas and lakes. We have a volcano, there are mountain ranges. There are dunes at the low latitudes within thirty degrees or so of the equator.'
Carrie Anderson, Cassini mission

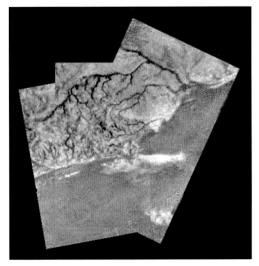

much more than half an hour's activity on the ground. For a mission that had cost millions of dollars and taken decades to plan and execute, its success would depend on this incredibly short window of exploration.

Despite its sophistication, Huygens could not communicate directly in any detail with Earth. Instead all of its detailed telemetry would need to be relayed back to Earth via Cassini's communications system. This meant very little was known about the probe's status during the 2-hour-and-27-minute descent, but at precisely 11.38 UTC, the Robert C. Byrd Green Bank Telescope in West Virginia picked up the weakest of carrier signals generated by Huygens's 10-watt internal transmitter to confirm that Huygens had arrived safely on the surface. This extraordinary probe was now alive and active on an alien moon orbiting Saturn.

Just under five hours later, the Cassini probe orbiting far above the moon's surface began relaying the precious data from Huygens back to Earth. And by 19.45 that evening the first extraordinary picture was released, showing a view from Huygens about 16 kilometres above the fast-approaching surface.

This was space exploration on an awesome scale: a photograph taken by a tiny space probe floating down on a parachute to the surface of a moon orbiting a planet over a billion kilometres from Earth. It was instantly apparent that this is no featureless rock: the channels running across this picture are recognisable as the geology of a world shaped by liquid, covered in drainage channels crossing the landscape into what looks like a lake bed or sea. And more was to rapidly follow. Just a few minutes later the first image from the surface arrived on Earth, the first view of the surface of a world in the frozen far reaches of the Solar System. It's an image that is worthy of the very closest of inspection, because, as always in science, the more you know about it the more wonderful it gets.

What we can see in this extraordinary image is something that is exactly like a floodplain or riverbed here on Earth. We can say this with some certainty because you can see that the rocks are abraded, they have been smoothed by the action of a liquid flowing over them. But there are no rocks. Huygens was able to confirm that these are in fact boulders of frozen water ice, sculpted on the freezing temperatures of Titan's surface. These rocks of ice are sitting on a surface that, as one Huygens scientist described it, is like 'snow that has been frozen on top'. 'If you walk carefully, you can walk as on a solid surface, but if you step on the snow a little too hard, you break in very deeply.' So if the temperature is so cold on Titan that rocks are made from water, what liquid is flowing over the rocks of ice carving the river beds and floodplains?

We'd known for some time from spectroscopy carried out here on Earth that Titan had an atmosphere rich in methane gas, and many scientists had wondered if this could mean that liquid methane existed on the surface. But it was only when Huygens was actually on the surface and its instruments began testing the atmosphere directly that we could begin to see just how alien this world was. For an unexpectedly long 70-minute period Huygens was active on the surface, detecting significant amounts of methane in the air. Huygens was also able to confirm that the surface temperature on Titan was a freezing minus 180 degrees Celsius; combined with an atmospheric pressure equivalent to 1.4 atmospheres here on Earth, the measurements confirmed that on Titan methane was not a flammable gas but potentially in abundance on the surface in liquid form.

Titan, it seemed, was a wet world, not with flowing water but with liquid methane driving ice-like rocks down mountain channels and out onto open floodplains. But despite early speculation, Huygens had not made a splash landing; it lay in a dry river bed and after surviving for just a few hours, its battery ran dead. This miraculous explorer switched off without directly detecting liquid methane on the surface. It would be two years later, as Huygens lay frozen and dead on the surface of the moon, that the very much alive Cassini probe would fly high above the south pole of Titan and beam back an image that would provide a unique view seen nowhere else in the Solar System beyond Earth.

This was our first glimpse of liquid on the surface of an alien world, a lake composed of methane, ethane and propane near the south pole of Titan. Named after one of the great lakes of North America, Ontario Lacus is like nothing we have ever seen on Earth, a 15,000-square-kilometre lake of hydrocarbons, whipped by winds that are clearly eroding the shoreline.

And Ontario Lacus was just the first. Today, thanks to Cassini's years of observation and photography, we have discovered more than 40 lakes of liquid methane, including Ligeia Mare, a vast body of liquid hydrocarbons with a shoreline over 200 kilometres in length. We've seen mysterious bubbles rising from its depths. Kraken Mare is the largest known body of liquid we have so far found on the surface

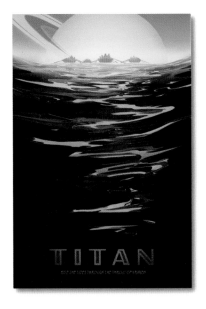

‘How vast those Orbs must be, and how inconsiderable this Earth, the Theatre upon which all our mighty Designs, all our Navigations, and all our Wars are transacted, is when compared to them. A very fit consideration, and matter of Reflection, for those Kings and Princes who sacrifice the Lives of so many People, only to flatter their Ambition in being Masters of some pitiful corner of this small Spot.’
Christiaan Huygens

of Titan; this hydrocarbon sea covers an area of 400,000 square kilometres – five times larger than Lake Superior – and has a depth that we suspect descends to 160 metres. We've seen waves in this sea and islands and currents, endlessly being filled by the flash floods of methane rain that we are certain fall on the highlands surrounding the sea feeding the rivers and streams that flow into it.

Earth, it seems, has a strange, cold twin, a moon not a planet that is millions of miles away, where lakes are liquid methane and mountains are made from frozen water ice as hard as rock. What's also fascinating, and tantalising, is that Titan has a complex chemistry, a carbon chemistry – the chemistry of life. We have found molecules like hydrogen cyanide, which are the building blocks of amino acids, molecules like vinyl cyanide, which chemists and biologists speculate could form some sort of cell membrane. In fact, all of the ingredients for life are present on the surface of Titan today. That doesn't mean there is life – in reality very few scientists think there will be life on Titan today, as it is, after all, minus 180 degrees Celsius at the surface. But with a little heat, it might be a very different story.

In a few billion years' time, as the light of the dying sun reaches out beyond the inner Solar System to touch the outer reaches with its warmth for the very first time, Titan will begin to warm. Its mountains of ice will shrink and melt and the frozen water they contain will replace the liquid methane, which will have evaporated away into space. Mountains will become oceans and in a strange twist of fate, at the end of the life of our star, perhaps the Solar System's last water world will be born. David Grinspoon believes it is a prediction we cannot rule out:

> ‘Titan today has all these juicy organics pooled and lying on its surface, and yet it's so cold and so dry that it may be that nothing is really happening biologically there. But we can imagine a time in the far future when the Sun has heated up to the point where the inner solar system has become uninhabitable, where Titan may become quite habitable. That methane greenhouse will become more powerful as the Sun gets hotter and at some point that icy crust will start to melt and you'll have large pools of liquid water on the surface full of all those complex organic chemicals, and that is a recipe for the kind of place where the origin of life can happen. So even if Titan doesn't have life today, and that's something we still have to search for, there's every reason to imagine that at some point in the future it could be a great place for biology.'

It's easy to think of habitability as a permanent feature of worlds – a defining characteristic, if you like. Earth is habitable because it's in the Goldilocks zone around the Sun – not too close and not too far away. But it's more complicated than that. Solar systems are dynamic places – over long timescales planets change orbits and their parent stars evolve, and worlds that were once heaven can become hell.

We now understand that Earth has been a fortunate world – an oasis in a constantly changing Solar System that's maintained a stable climate, perhaps against the odds, for the 4 billion years it's taken complex living things to evolve. We don't know how many planets like Earth are out there amongst the stars – where the ingredients of solar systems have assembled into structures capable of dreaming of other worlds – but we must consider the possibility that there are very few, and that would make Earth – and us – extremely rare and precious.

Opposite top: Huygens returned images of what appear to be pebble-sized rocks or ice blocks across the surface of Titan.

Opposite bottom: Just one of the many lakes of liquid methane discovered on Titan, these images of Ligeia Mare hint at the presence of waves.

ENCELADUS

Nearly a billion miles from the warmth of the Sun, on the outer edge of Saturn's rings, lies the icy moon of Enceladus. A small moon, about the size of Iceland, Enceladus is the most reflective object in the Solar System – over 90 per cent of the light that hits it bounces back. That's because the surface of this ice-covered moon is covered by crevasses just like we see on the surface of glaciers here on Earth, but on Enceladus these cracks in the ice occur on a much grander scale, with some stretching for over 100 kilometres. But for all that's going on on the surface, it's what we discovered beneath Enceladus's shell of ice that must rank as one of the greatest surprises in twenty-first-century space exploration. A surprise that had been hiding in plain sight for decades.

The first photograph of Enceladus was taken by Voyager 1 in 1980. It was actually captured unintentionally, a moon photo-bombing the planet, and nobody even noticed Enceladus until over 35 years later. But when the image was eventually enhanced, a remarkable feature, no more than a misty blob, began to emerge on one side of the moon. Twenty-four years later, Cassini saw exactly the same thing, but this time in much more detail. As Cassini approached Enceladus, the anomaly revealed itself. Giant plumes of water vapour and ice were erupting from its surface; over 200 kilograms of material was being released every second, feeding one of Saturn's outer rings and helping to replenish it.

These discoveries inspired the Cassini team to take an audacious risk. Piloted from nearly a billion miles away, Cassini was guided dangerously close to the plumes and using its Cosmic Dust Analyser was able to actually touch the plumes of ice and

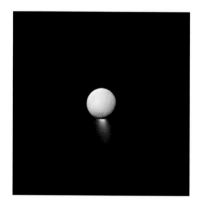

Above: Enceladus's rarely seen plume, captured by Cassini. Its light is reflected off Saturn, enabling this moment to be recorded.

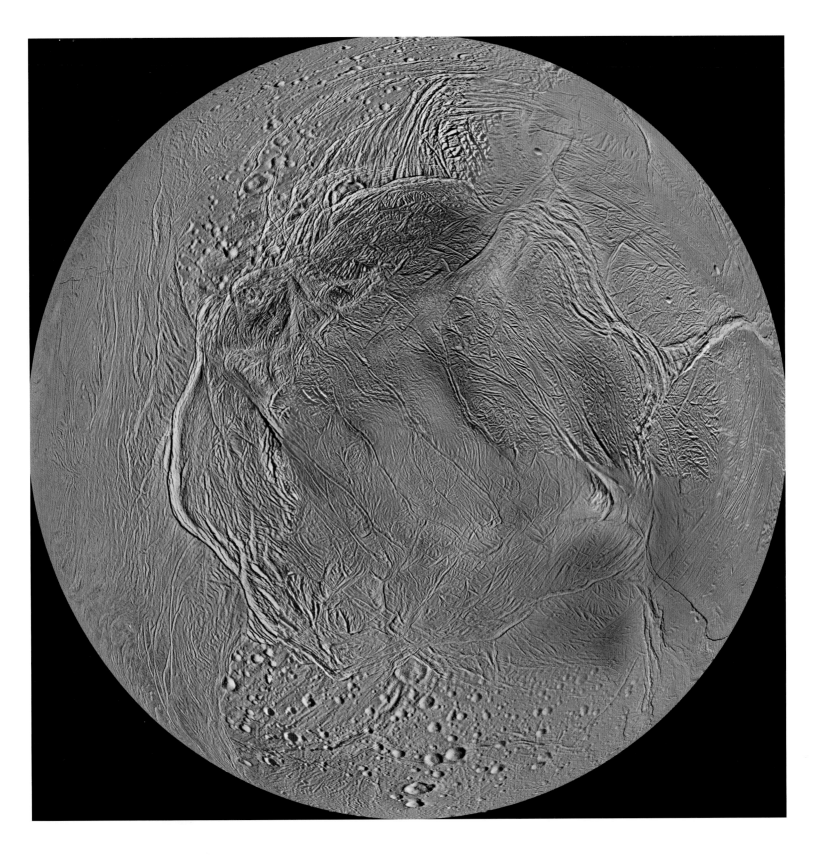

Opposite: Our exploration of our icy outreaches on Earth may give us an insight as to whether there is any life on frozen, icy Enceladus.

Above: This illustration marks out the locations of potential origins of methane in the gas and ice particles that Enceladus sprays out.

Below: Hydrothermal vents on Enceladus release methane and other potentially life-creating molecules from the planet's interior.

Opposite: Scientists have been able to model eruptions on Enceladus along fractures across the moon's south pole.

collect a sample from one of the eruptions. As Michele Dougherty, scientific leader on the Cassini Project, explained, 'The really close flyby that we had of Enceladus where we were 25 kilometres above the surface is probably closer than any spacecraft has been to a planetary body before, and the Cassini mission team said they'd never do it again because we were in the atmosphere, well in the plume which was really dense. They didn't lose control of the spacecraft but they felt they might because the mag boom used to stickle from the side of the spacecraft and that almost started tumbling the spacecraft.' What the scientists discovered, though, thanks to this and a series of similar flybys, was breathtaking. The plumes were spewing out of fissures, known as 'tiger stripes', being projected from deep within this ice world, rising up from an ocean of salty liquid water beneath the ice.

How could such an ocean exist in the far frozen reaches of the Solar System? The answer lies once again in the power of Saturn to control the destiny of the moons in its dominion. Unlike the long-lost moon of Veritas that was perhaps destroyed by the might of Saturn to create its rings, in the case of Enceladus its mother planet has breathed life, not death, into its structure.

ENCELADUS VENTS

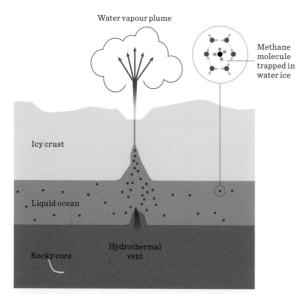

Water vapour plume

Methane molecule trapped in water ice

Icy crust

Liquid ocean

Rocky core

Hydrothermal vent

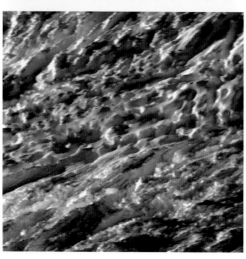

Left: In its closest dive towards Enceladus, Cassini captured this image of the moon's surface.

Above: Pale Dione and Enceladus (behind) are composed of similar materials, yet Enceladus's reflectivity makes it appear brighter in the sky.

Opposite: Black smoker hydrothermal vent along the Juan de Fuca Ridge in the Pacific Ocean. Life on Earth likely began in vents like this, and those found on Enceladus.

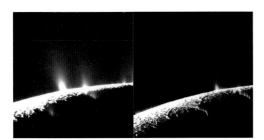

Enceladus orbits Saturn and the planet's vast gravitational force pulls at the moon, holding it in orbit, but every couple of rotations another larger moon called Dione aligns with Enceladus, drawing it back. This repeated tug of opposing tidal forces stretches and squeezes the moon, warming its core and melting its icy interior. Exactly the same gravitational forces that destroyed one of Saturn's moons are now breathing life into another. But it's what Cassini found in the chemical makeup of the plumes that changed everything, because as it analysed the columns of water, it discovered complex organic compounds and silica particles, direct evidence that deep below the ice at the bottom of a salty ocean of water there were hot hydrothermal vents exactly like the ones from which we believe life emerged on Earth almost 4 billion years ago.

Out in the frozen outer reaches of the Solar System, Cassini has given us a glimpse of the world beneath the ice on Enceladus and revealed a warm, watery oasis, a world in which it really is possible that life could exist. Exciting though the prospect of life in space is, if it is out there on Enceladus it's likely only to be home to the simplest and most primitive organisms, and given how violent and changeable Saturn's past

has been, this world may have arisen relatively recently. We don't know how long Enceladus has been active, but if it's only tens or even hundreds of millions of years, that may not have been enough time for life to get going. And yet despite all the caveats, the very fact that the Saturnian system holds the potential to host life is a truly remarkable discovery, and one that compels us to explore even more.

But for Cassini, the brave new world under the ice of Enceladus was a bittersweet discovery. Thirteen years after it arrived in orbit around Saturn, as Cassini's fuel finally began to run dry, the team at NASA realised it couldn't risk letting this probe crash-land on Enceladus and contaminate a potential habitat for life. As Carl Murray explains, 'Because we now knew Enceladus was a potentially habitable environment, we couldn't risk Cassini crashing into it or coming near it, because there might be microbes hitchhiking on the spacecraft, and we couldn't allow contamination of Enceladus. So we consciously picked an end of mission that would use up its fuel and then send the spacecraft into the giant planet itself.'

We journeyed out across space to Saturn to see this great beauty up close, but in the end we were rewarded with insights that hit far closer to home. As is so often the case in exploration and science, the real treasure was found in the unexpected shadows – on this mission, it was in the shadow of the rings, because there we found a tiny world with the potential to harbour life, to provide a hint of reassurance that for the first time we might not be alone.

Would such a discovery matter? Yes, it would – at this moment in history we are, I think, in need of reassurance, or perhaps a reminder, that there is great beauty and knowledge and perhaps even life and meaning, beyond our shores. We will, I'm sure, go back to Saturn and its moons, explore deeper, stay longer and maybe one day even visit there ourselves. We don't yet know what we will discover, but we can be sure it's a story that has only just begun.

'Cassini's discoveries have profoundly changed our understanding of the Saturn system, from the planet itself to detailed structure in the rings. We now know what the surface of Titan looks like, and we have also pierced through that haze with the radar to map it all out. The lakes and seas are liquid methane on the surface of this moon.

And then of course the big discovery of geysers coming out of the south pole of Enceladus, flying through those geysers, sampling, tasting, measuring what's coming out, the gases and the particles, and finding not only do you have a lot of water from that liquid water ocean but you have organics. You have nitrogen, you have methane also, so you have sort of the key ingredients, carbon, hydrogen, oxygen, nitrogen – the building blocks for life.

It's so intriguing to find a moon that's only 500 kilometres across with a liquid water ocean, just one of a suite of ocean worlds in our Solar System, and perhaps life is just next door.'
Linda Spilker, Cassini mission

Opposite: The moon Enceladus setting behind its planet Saturn. This image was taken in the last days of Cassini's Grand Finale.

Above: A giant of a moon appears before a giant of a planet in this natural color view of Titan and Saturn from NASA's Cassini spacecraft.

URANUS NEPTUNE PLUTO

INTO THE DARKNESS

ANDREW COHEN

'In order for the light to shine so brightly,
the darkness must be present.'
Francis Bacon

'This is a historic time; future generations
will look back and say "this is where we
emerged from Earth's cradle".'
Alan Stern

JOURNEY TO THE EDGE

Out there far into the darkness lie the most distant planets of the Solar System, worlds we can hardly even glimpse from Earth.

In a little over six decades, we've met every one of the planets in our Solar System. Travelling at first to our nearest neighbours, we visited Venus, Mars and Mercury and found rocky planets covered with the remains of long-lost worlds – each has a life that followed very different paths. To these planets we returned time and again, staying longer on each visit, exploring deeper and even touching down onto alien ground. Then we ventured further – beyond the rocky worlds, beyond the ice line and out to the land of the giants. As we explored Jupiter and Saturn, we began to piece together the extraordinary life story of these gas giants and all the exotic moons in their grasp.

But that was just the beginning. Beyond Saturn, out there far into the darkness lie the most distant planets of the Solar System, worlds we can hardly even glimpse from Earth: Uranus, a pale blue marble hanging in the frozen depths of space; further out its icy sibling Neptune; and beyond that Pluto, a world but not a planet, which until now has remained virtually hidden from us in all but name. Each lies not millions but billions of kilometres from Earth, and for this reason only once have we dared to visit them and capture a set of images from the dark – precious data that allows us to begin to understand the endless mysteries that lie out there at the outer edges of our Solar System and beyond.

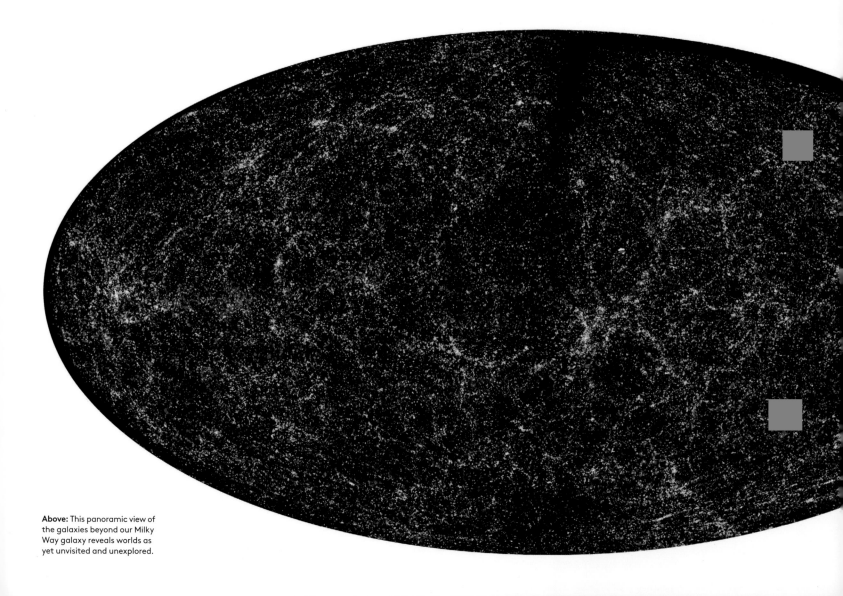

Above: This panoramic view of the galaxies beyond our Milky Way galaxy reveals worlds as yet unvisited and unexplored.

The scale of the Solar System is virtually unimaginable; a landscape laid out across distances impossible to comprehend within the realms of human experience. London to New York is 5,500 kilometres, London to Sydney a mere 17,000 kilometres, while London to the Moon is 384,000 kilometres, and the longest single journey any human has ever completed – the equivalent of travelling nine and a half times around the circumference of Earth. Beyond these journeys, as we look out towards ever more distant destinations, the numbers grow so big so quickly that we have had to invent a different unit of distance altogether, in order to give us a containable scale for the vastness of the cosmos. It's based on the mean distance between the Earth and Sun, which is 149,598,000 kilometres – known as 1 astronomical unit, or 1 au.

Today we can measure this distance with relative ease, but calculating that number for the first time with any accuracy was far from an easy task. For near on 2,000 years, many a great mind took measurements, made calculations and came up with numbers that estimated the distance to the Sun, but it wasn't until the seventeenth century that we were able to measure for the first time the distance to our star, when Joseph Kepler provided the mathematical underpinning and Edmond Halley the understanding that the transit of a planet could unlock our ability to accurately calculate an astronomical unit.

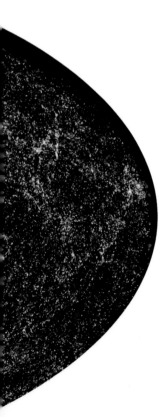

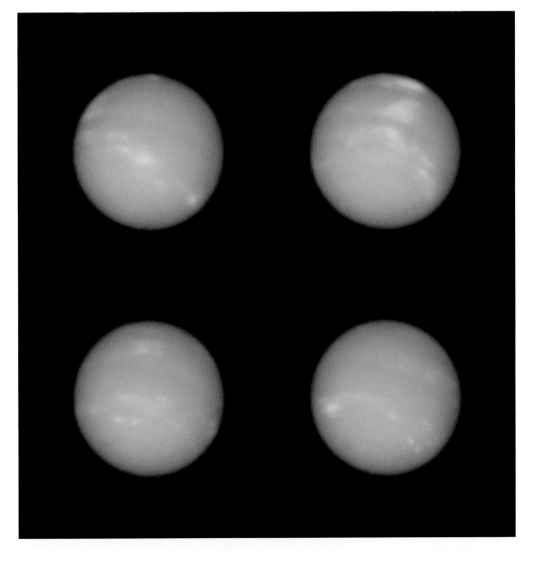

Above: It is thanks to Edmond Halley, who laid the groundwork for later scientists, that we can calculate distances across the Solar System.

Right: On 25 June 2011, Neptune arrived at the same location in space where it was discovered 165 years earlier. To commemorate the event, NASA's Hubble Space Telescope took 'anniversary pictures' of the blue-green giant planet.

Halley was the first to realise, in 1716, that by measuring a transit of Venus – the moment our sister world passes directly between us and the Sun – from different positions around the globe, we could obtain the necessary numbers to allow the geometry to play out. Unfortunately for Halley, he had the right idea at the wrong time, as transits of Venus are enormously rare, coming in pairs approximately once every 120 years. So in one of the great scientific baton-passes of all time, Halley left behind detailed instructions for future astronomers to carry out his plan. Twenty years after his death, the 1761 transit of Venus provided the first opportunity to put his theory to the test. In a remarkable combination of collaboration and competition, which occurred in the middle of the Anglo/French Seven Years War, observers from both these countries travelled all over the world to destinations as far afield as Siberia, Madagascar, Newfoundland and the Cape of Good Hope, praying for a clear sky to observe and measure the transit. Combined with a similar set of expeditions for the next transit eight years later, enough data was gathered to allow the French astronomer Jérôme Lalande to calculate the distance between the Earth and the Sun for the very first time, coming up with a pretty accurate figure of 153 million kilometres.

Today we no longer need to rely on the rarest of astronomical events to make these measurements. Instead we have far more direct ways of measuring that have allowed us to plot our journey around the Sun in ever greater detail. Throughout a year, as we follow our elliptical orbit around the Sun, the distance from our planet to the centre of the Solar System varies enormously. This means that in early January each year, as we reach the closest proximity to our star – known as perihelion – we are about 147 million kilometres distant. Six months later, in early July, we reach aphelion – our furthest distance from the Sun – and find ourselves about 5 million kilometres further away, orbiting at around 152 million kilometres away.

Exactly what one astronomical unit actually is has therefore been the subject of intense debate and change over the last century, and as we have sent our robotic explorers across the Solar System our ability to directly measure the precise distances between the inner planets using radar and telemetry has yielded endless refinements. So since 2012, we – or the International Astronomical Union on our behalf – have made life much more straightforward by defining 1 au as 149,597,870,700 metres or about 150 million kilometres.

Mars, the furthest out of the four rocky planets, sits at 1.5 au, half as far away from the Sun again as we are. At 2.2 au the asteroid belt begins; a swirling carousel of rocks and failed worlds, it stretches out for another 150 million kilometres before coming to an end. Here, at the snow line of our Solar System, 3.2 au away from the Sun, we leave the rocky worlds behind and continue across another 300 million kilometres of space before we reach the first of the giants – Jupiter. And yet here at 5.2 au, 780 million kilometres from the Sun, our journey has only just begun. It is almost double the distance again before we reach the next planet, the ringed jewel that is Saturn, sitting 9.6 au or 1,434 million kilometres from the centre. It is here that the realm of our exploration ends, at the last of the planets that we can say we have truly explored.

At 2 au the asteroid belt begins; a swirling carousel of rocks and failed worlds.

Opposite: Voyager 2 – the only spacecraft to have made flybys of all four outer planets – Jupiter, Saturn, Uranus and Neptune.

Above: One of the rarest predictable solar events: the transit of Venus. This image was taken in 2012, with the next transit due in 2117.

237

Bottom: The massive scale of the Solar System – from its fiery heart to the mysterious frozen Kuiper Belt at its outer border.

URANUS + NEPTUNE + PLUTO

Mercury, Venus, Mars, Jupiter, Saturn – each one a world that we have not only visited but resided around, our robotic explorers orbiting with camera eyes wide open for years at a time. In many cases we've not just looked at but also touched the surface of these worlds and the moons they embrace. Once we get beyond Saturn, the scales change so much, the distances grow so vast and travel times so long they are measured not in days and months but years and decades.

Uranus, the first of the ice giants, sits over twice as far away from the Sun as Saturn, orbiting at an average distance of 19.2 au or 2.871 billion kilometres. Then Neptune, the farthest of the planets, lies another 11 au further out, and orbits nearly 4.5 billion kilometres away from the Sun. After it comes Pluto, on the highly elliptical orbit that takes it inside and outside the line of Neptune but on average sits 5.9 billion kilometres away from the Sun. To reach these worlds is so difficult that in our 60 years of planetary exploration we've visited each only once, and spent just the briefest of moments in close proximity to them. We manage a matter of hours in each system, as our spacecraft are propelled at enormous speeds, so fast that they quickly fly past and beyond. But the precious scraps of data they have returned have allowed us to peer through the mysteries that lie out there at the outer edge of our Solar System and start to tell the extraordinary stories that reach deep into the darkness.

41 years, 4 months, 19 days and counting

To reach the outer planets has taken the work of perhaps the greatest of all of our planetary explorers. At the time of writing, in January 2019, the Voyager 2 spacecraft has been in continual operation for 41 years, 4 months and 19 days. Travelling at 15.341 km/s, it is currently around 119 au from the Sun – that's 17 billion kilometres away – and has just recently been officially deemed to have entered interstellar space. No longer within the empire of our Sun, the heliosphere created by the solar wind that defines the boundary of our Solar System, this little probe weighing less than a family car is the third human-made object after its sibling Voyager 1 and also Pioneer 10 to leave our home star system behind. As it heads out into the void it is expected to continue to stay in contact with us here on Earth for at least another six or seven years, transmitting its weak radio signals until at least 2025, before slipping off into its final silence around 48 years after its launch, on its way towards a small star called Ross 248, located in the northern constellation of Andromeda.

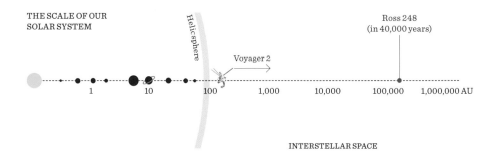

THE SCALE OF OUR
SOLAR SYSTEM

Heliosphere

Voyager 2

Ross 248
(in 40,000 years)

1   10   100   1,000   10,000   100,000   1,000,000 AU

INTERSTELLAR SPACE

THE GRAND TOUR

'If you launch from Earth at just the right time you can fly by Jupiter in such a way that you can use its gravity to fling you onto Saturn, then hit Saturn just right to fling you onto Uranus and Neptune. That notion was called the Grand Tour … and that plan became the Voyager mission.'
David Grinspoon, astrobiologist

It all began on 20 August 1977, when the Voyager 2 spacecraft left our planet on board a Titan III rocket, taking off from Launch Complex 41, Cape Canaveral Air Force Base. Two weeks later, on 5 September, the twin probe Voyager 1 would follow hot on the heels of its sibling. Taking the fast lane, Voyager 1 would speed ahead, reaching Jupiter four months earlier than Voyager 2, which had been sent on a different trajectory that took it on a slower, longer, more circular course.

Making its closest approach on 9 July 1979, Voyager 2 swept past Jupiter at a distance of 576,000 kilometres from the planet's cloud tops, gathering a vast amount of data on the giant planet's atmospheric characteristics and its moons. This included the first detailed analysis of the swirling storm known as the Great Red Spot, as well as a ten-hour analysis of the volcanoes of Io, allowing the activity to be calculated by directly comparing the images of the moon with images Voyager 1 had taken just a few months before.

Using Jupiter as a gravitational springboard, Voyager 2 then left the Jovian system behind and continued on a two-year journey to Saturn. With its closest approach occurring on 26 August 1981, Voyager 2 became the third spacecraft to visit Saturn, providing a vast range of pictures and data on the planet, its rings and moons. But it was the next leg of its journey that would be truly groundbreaking, and it was only possible because of the rarest of celestial alignments.

Back in 1964, Gary Flando, an aerospace engineer working at the Jet Propulsion Laboratory in Pasadena, California, noticed that an unusual configuration of the four outer planets would occur in the late 1970s and early 80s. Flando realised that this particular alignment, which occurs once every 175 years, would potentially allow a single spacecraft to visit all the outer planets using a series of flybys and gravity assists that would provide an opportunity to visit not just Jupiter and Saturn but Uranus and Neptune for the very first time.

Over the next 15 years NASA began plotting a mission to complete what became known as the Grand Tour. Various plans for the Tour were started and shelved as the costs and feasibility of the proposed route came into question, until eventually in 1972 the twin probe Mariner Jupiter-Saturn project was approved with an estimated cost of $360 million. The two probes would eventually be renamed the Voyagers, and only one would complete the Grand Tour. Voyager 1 would be sent from Jupiter to Saturn and then on to its satellite Titan to attempt to explore the moon and its intriguing atmosphere, but Voyager 2 would be given the grandest of journeys and the honour of becoming the first craft to attempt to see Uranus and Neptune close up.

Flying past the Saturnian system in late August 1981, Voyager 2 began a journey into the darkness that has yet to be repeated. Over the next three and a half years the craft would make its lonely journey outwards towards its unique encounter with the third-largest planet in the Solar System, a planet that until then we had only ever been able to peer at in the distance.

Opposite: On 20 August 1977 Voyager 2 launched, beginning a groundbreaking journey of discovery.

THE PALE BLUE MARBLE

The seventh rock from the Sun sits 2.9 billion kilometres from the Earth, a pale blue marble hanging faintly in our night sky. Barely visible with the naked eye, ancient observers dismissed the object as a star – too slow and too dim to join the family of planets. It wasn't until the invention of the telescope that the real character of this distant object began to emerge, but even then the discovery of Uranus was a slow one.

It began on a cold March night in 1781 – Tuesday 13th, between 10 and 11 pm, to be precise – in the townhouse garden of William Herschel. As the 43-year-old astronomer peered through his 160-mm telescope, surveying the sky that night for any small stars he could bring into view, he noticed a dim object that over the course of the next few days appeared to be slowly moving, so he assumed it must be a comet. Reporting his findings to the Royal Society the following month, Herschel still seemed convinced this was a comet, not a planet, even though he himself suggested that the characteristics of the observation were 'as planets are'. The Astronomer Royal at the time, Nevil Maskelyne (whose claim to fame is being the first person to

Right: Artwork showing the Voyager 2 probe heading towards its closest approach of Uranus, which lasted six hours.

'By the observations of the most eminent astronomers in Europe it appears that the new star, which I had the honour of pointing out to them in March 1781, is a Primary Planet of our Solar System.'
Sir William Herschel, 1783

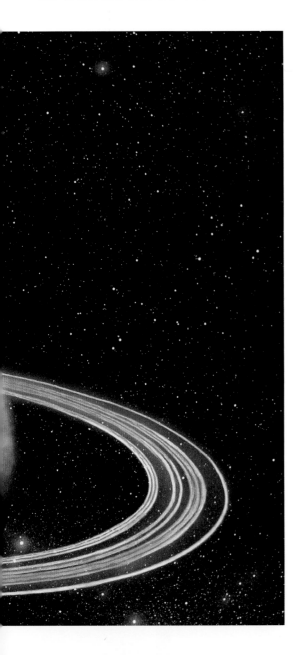

measure the weight of the Earth scientifically for the first time), wrote to Herschel in late April 1781 to share his bafflement. 'I don't know what to call it,' he wrote. 'It is as likely to be a regular planet moving in an orbit nearly circular to the Sun as a Comet moving in a very eccentric ellipsis.'

Eventually the accurate computation of the body's orbit calculated by a number of different astronomers provided the ultimate proof of its character, and its planetary status was confirmed by Herschel in a letter to the President of the Royal Society, Sir Joseph Banks, in 1783.

'Sir – By the observations of the most eminent astronomers in Europe it appears that the new star, which I had the honour of pointing out to them in March 1781, is a Primary Planet of our Solar System.'

With that, the seventh planet was born into our story of the Solar System, but naming this new arrival – the first planet to require such a decision since antiquity – was far from straightforward. The announcement of Herschel's discovery had not only impressed the world of science but also captured the attention of the higher reaches of British society, all the way up to King George III. As reward for his historic achievement, Herschel came under the patronage of the king to the tune of £200 a year, and moved to Windsor to allow the royal family the use of his telescopes. Herschel had the perfect response in mind to return such favours. Charged with the responsibility of naming the new planet, he proposed the name Georgium Sidus – 'George's Star' – after the king himself. However, the name met with immediate opposition away from British shores, so other more universal alternatives were proposed, including Herschel, Neptune and Uranus – after the Greek god and father of Saturn. Amazingly, Georgium Sidus continued in some usage for the next 70 years, but by 1850, when His Majesty's Nautical Almanac Office finally caved in, the name was switched to Uranus, and it finally and permanently stuck.

It would be 136 years later, across 2.6 billion kilometres of space, when the planet that was discovered in the back garden of a Somerset town house would slowly expand into the viewfinder of Voyager 2. Approaching it at 17 km/s, after a journey of over eight and a half years, our first and only close encounter with this most enigmatic of planets would play out within a matter of just six or so hours.

Remarkably little was known about Uranus, as the spacecraft began the observatory phase of its mission in early November 1985. We knew this world was the third-largest by volume after Jupiter and Saturn and that it took a leisurely 84 years to complete a single highly elliptical orbit, but we could only estimate the length of its day to be somewhere between 16 and 24 hours. We also knew its composition was substantially different to the two gas giants with higher amounts of water, ammonia and methane compared with its bigger neighbours, but how this formed an atmosphere and what dynamics played out beyond the pale blue mist was a mystery. We also knew that it had at least five moons, tiny points of light in orbit that we had detected with telescopes from Earth, but again, what form these moons took and what features they held remained a complete mystery, as was a faint and narrow set of rings around the planet that had only been discovered a few years earlier.

Above: Uranus was the first planet discovered with the aid of a telescope – William Herschel's reflecting telescope, to be precise.

'There really are no other credible theories out there for why Uranus's rotation axis should be tilted, other than a giant impact. It sounds a little bit of a cheat, because when in doubt invoke an impact, but that's usually the answer when there's an odd feature for a planet. So something the size of two, three even five Earth masses is what is required to have hit Uranus to knock it on its side.'
Steve Desch, astrobiologist

What these rings were made of, and why they had formed from such dark material, compared with the glowing rings of Saturn, was yet another element that needed exploration, but perhaps most mysterious of all was the fact that the whole Uranian system had the strangest of orientations. The planet's rotation, its moons and its rings all appeared to be tipped up on the side, rotating around a horizontal rather than vertical axis, in contrast with every other planet in the Solar System. As astrobiologist David Grinspoon explains, 'The Uranian system really stands out in the Solar System in several ways. But the most striking way visually is that it's tilted on its side, almost completely, so that the north pole of Uranus sometimes points right at the Sun, and it's always in that plain of the orbits, and the moons as well, are orbiting in that extremely tilted plain, so the whole system forms this sort of bullseye pattern that is rotating at this odd angle around the Sun.'

At 9.59 PST on 24 January 1986, the spacecraft flew just over 81,500 kilometres above the cloud tops of Uranus at its closest proximity to the target. Back on Earth, receiving stations looked up to the sky impatiently, pointing their giant antennae towards the far explorer in anticipation of the first of its faint radio signals. But for all the anticipation and expectation of a mission that had been decades in the planning, years in transit and with a catalogue of extraordinary revelations already behind it, the first impressions were perhaps a little underwhelming.

The first images returned by Voyager revealed a beautiful but featureless world, a pale blue orb hanging in the frozen reaches of our Solar System that was covered in an unbroken haze, in striking contrast with the dynamic realms of Jupiter and Saturn that Voyager had revealed to us in such detailed beauty. Peering into the haze, only ten cloud features could be seen across the whole planet, blown around by winds travelling in the same direction as the planet rotated. But despite first appearances, slowly but surely the data that Voyager returned would change our understanding of almost every aspect of this world.

Voyager was able to measure the rotation of the interior of the planet for the first time, revealing that a 'day' on Uranus lasted for 17 hours and 14 minutes.

Beyond that, it also revealed Uranus to be a bitterly cold world, with the average temperature being an icy minus 213 degrees Celsius. In fact, we now know that Uranus is the coldest planet in the Solar System, and with such cold comes a stillness, the lack of internal heat helping to explain the inactivity of its atmosphere. Why Uranus's internal heat appears to be so much lower than that of the other giant planets is still poorly understood.

As Voyager sped past it also attempted to measure the composition and structure of Uranus. As expected, the most abundant ingredients in its atmosphere were found to be hydrogen (83 per cent) and helium (15 per cent), although the amount of helium was lower than many scientists had predicted. The relatively high level of methane measured in the upper atmosphere, at 2.5 per cent, also provided an explanation for the aquamarine-blue glow of the planet that we'd seen from afar for so long. But understanding more about the planet's internal structure was not quite so straightforward.

ANATOMY OF URANUS

50,724 km

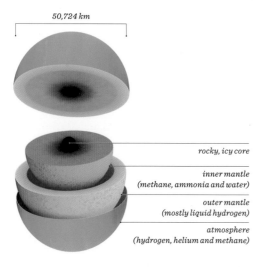

rocky, icy core

inner mantle
(methane, ammonia and water)

outer mantle
(mostly liquid hydrogen)

atmosphere
(hydrogen, helium and methane)

243

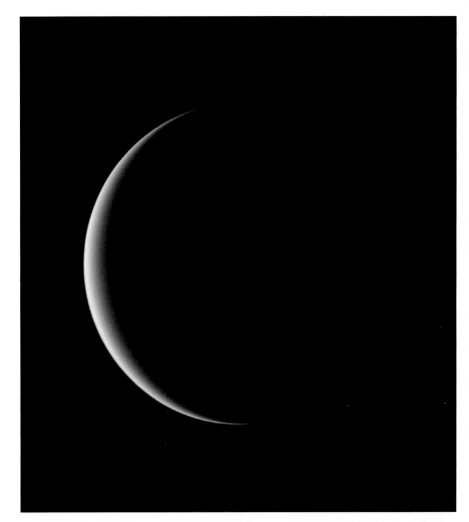

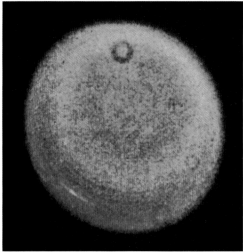

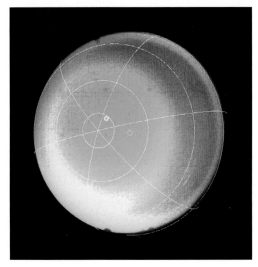

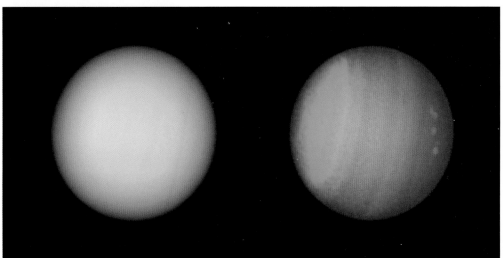

Above: The latitude-longitude grid superimposed on this image of Uranus's atmosphere shows it circulates in the same direction in which the planet rotates.

Above: Two images of Uranus; on the left it appears in its true colours; the right image was taken using filters to show cloud bands in its atmosphere.

<u>David Grinspoon on Uranus's diamond rain</u>
*'Uranus is an oddball even in a solar system of oddballs.
It's the closest of what we call the ice giants, which
means it's not made out of primarily gases, like Jupiter
and Saturn, but it's not made out of primarily rock like
Earth and the other terrestrial planets. It's sort of,
in-between, so it has a giant gaseous envelope, but a
core of rock and ice.*

*It's huge – about four times the diameter of Earth
and about 15 times the mass of Earth – and it seems
so simple on the surface, but we know in the interior
there's a lot going on. There are clouds of different
materials – water, methane, we think – in the interior.
It may be raining diamonds, so that featureless,
simple, bland exterior hides a very complex interior
of all kinds of interesting chemistry and physics.'*

At roughly 14 and a half times the mass of the Earth, Uranus consists mostly of water, ammonia and methane in the form of 'ices' that we believe are wrapped around a rocky core. However, the exact masses that make up each of these layers and the amount of each of the individual constituents was beyond the measuring capabilities of Voyager, so even today we can still only estimate the exact makeup of the planet.

However, the data returned by Voyager did result in one major change in our understanding of the planet's interior. With the knowledge that the composition of Uranus (and soon Neptune) was substantially different from the gas giants of Jupiter and Saturn, a new category of planet was defined in the 1990s from the data the spacecraft gathered. Uranus is one of two ice giants in our Solar System, a type of planet distinct from the gas giants of Jupiter and Saturn that formed in the warmer region nearer to the Sun.

Above: Scientists have carried
out high-pressure experiments
to conclude that it might in
fact rain diamonds on Uranus.

Above: Uranus as imaged by
Voyager 2 in 1986; its blue-green
colour comes from the methane
gas in its atmosphere.

Opposite: One of two ice
giants in the outer Solar
System – the other being
Neptune – Uranus, unlike
Earth with its icy, glacial
outposts, is covered with
swirling 'icy' fluids.

We still understand relatively little about the formation of the ice giants beyond the very basic facts. What we do know is that around 4.5 billion years ago, as each of the planets was forming from the ingredients of the proto-planetary disc, Uranus found itself in a region where there was an abundance of water, methane and ammonia at temperatures where these volatile compounds (each with freezing points above 100K, or minus 173 degrees Celsius) were cold enough to form the 'ices' from which most of the planet would be built. Beyond this we can tell relatively little about the mechanism by which either Uranus or Neptune actually formed. The models that have allowed us to tell the earliest beginnings of the Solar System and revealed so much about the formation of the gas giants and the inner planets do not yet exist for the ice giants. We simply lack a solid understanding of the mechanism by which these planets formed, and much controversy remains around the details.

URANUS'S RINGS AND MOONS

While there remain many mysteries out there in the darkness, Voyager did allow us to begin to piece together one of the most dramatic stories that unfolded after the formation of Uranus. As the craft sped past the planet's pale blue atmosphere it began to explore one of the least-known gems in the Solar System, the rings surrounding this most distant of worlds. Unlike the shimmering rings of Saturn, these rings are so dark and faint that they are almost impossible to see from Earth. In fact, we only discovered them by accident less than a decade before Voyager arrived in the Uranian system, when in 1977 a team from the Kuiper Airborne Observatory noticed a star that was briefly disappearing from view around the planet. The conclusion they came to was that the star was being occluded by a ring system. At first they confirmed five rings, but they increased this number to nine after further observation.

Voyager was able to image all nine of the rings in the system, sending back extraordinary images of these dark delicate structures and also discovering two other faint rings in the process. But as well as giving us our first detailed view of the rings, Voyager added to their mystery. Made up of a combination of boulders and dust, the reflectivity of the substance of the rings as measured by Voyager was found to be incredibly low. The reason why these rings were so difficult to detect from Earth was that unlike the glistening ice that dazzles in the rings of Saturn, the rings of Uranus turned out to be grey and made of an extremely dark material, making the spectral analysis of them extremely difficult, if not impossible. Despite the spacecraft's close proximity, we were unable to make any measurements of the constituents of the rings, and over thirty years after our only close encounter we still don't know precisely what they are made of or where they came from. Our best guess is that they are made of ice and some other dark substance, perhaps even organic compounds that have darkened over time. The most likely scenario is that just as with the rings of Saturn, Uranus is surrounded by the debris of a moon, a world that once orbited her and that is now broken into the millions of fragments of the rings.

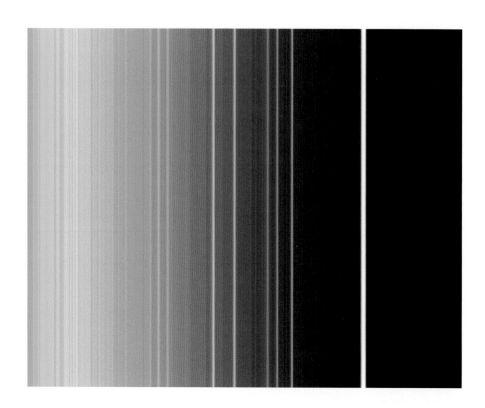

Above and right: In its historic flyby, Voyager 2 took this picture and false-colour view which shows nine of Uranus's rings. The most prominent is epsilon, with three more – delta, gamma and eta — appearing much fainter and narrower.

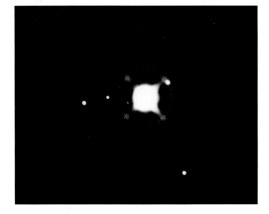

'We knew the rings were there at Uranus, but we didn't know their configuration or their shape, they were very dark, they were very thin and in orbit around Uranus, where they are shepherded and made into thin rings by interactions with the moons around them.'
Fran Bagenal, New Horizons mission

The rings are also extremely delicate and thin, an ephemeral set of structures that appear to defy the laws of nature. We would expect the particles to endlessly collide and the rings to spread out and begin to disperse over time, a process that should be driven by the particles on the inside of the rings being drawn into the planet by its great mass. But that doesn't seem to be happening: something seems to be holding the rings in place. Hidden in the images taken by Voyager 2, we found a surprising answer.

The outer ring, epsilon, is the brightest and densest of all the structures in the system and its width varies from 20 kilometres at its narrowest to just under 100 kilometres at its widest. This fragile structure is also thought to be unimaginably thin – we have never been able to measure its depth precisely but we think it could be little more than 150 metres deep. Maintaining this delicate structure in the hostile environment around the giant planet is the job of two tiny moons, Cordelia and Ophelia, whose existence was first discovered from images taken by Voyager 2. With a radius of just 21 kilometres, Ophelia sits on the outside of the ring, while the slightly smaller Cordelia, the innermost known moon of Uranus, sits on the inside of the epsilon ring. Together these two moons exert a profound influence on the structure and stability of the ring they embrace. These 'shepherd' moons quite literally guide the movement of objects around the edge of the ring, helping to maintain the sharply defined border that we've seen so vividly from Voyager's images.

To understand the action of the shepherd moons we need to look at the physics of the system. Simple orbital mechanics dictate that a moon or a ring particle that is orbiting close to the planet is moving more slowly than a moon or a ring particle that's orbiting further away, which means that Ophelia must be moving faster than the ring particles and Cordelia more slowly. That difference in speed is significant, because if there's a particle in the epsilon ring that shifts for any reason – maybe it has a collision, loses a bit of energy and starts to drop down – then Cordelia will tend to accelerate that particle back up into the ring again; speeding it up by its gravitational influence. Whereas if there's a particle on the outer edge of the epsilon ring that gets a bit more energy and rises in altitude, Ophelia tends to slow it down, and so it falls back into the ring again. It's by this mechanism that these shepherd moons keep the ring neat and tidy, and although Voyager only discovered two such moons on its brief flyby of the system, we can assume from the structure of the other rings that somewhere out there in the dark there are other hidden shepherds, silently keeping order amongst the rocks and dust of their flock.

These two are just the beginning of Uranus's system of satellites. As of today, we have discovered 27 moons around Uranus, all of which are named after characters from the works of Shakespeare or Alexander Pope. This tradition began in 1852, when John Herschel – the son of William, discoverer of Uranus – named the first two moons after Oberon and Titania, fairies from *A Midsummer Night's Dream*.

The 27 moons are divided into three categories; 13 inner moons, including the two innermost – Cordelia and Orphelia – are all small dark bodies that are entwined within the ring system and may even share an origin with the substance of the rings. Then much further out lie nine irregular moons, ranging in size from 120 to 200 kilometres in diameter; these are almost certainly captured objects that have succumbed to the attraction of the planet's gravity.

Top and bottom: Voyager 2 discovered two satellites – Ophelia (1986U8) and Cordelia (1986U7) – associated with the rings of Uranus.

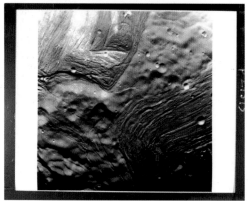

In between this rabble of objects and inner moons we find the five major moons of Uranus, the only Uranian satellites we knew of before Voyager gave us a closer look. The smallest of the set is the ice moon Miranda. A strange-looking, jagged world, with a diameter of 470 kilometres, it's one of the smallest objects we have studied in the Solar System that has reached hydrostatic equilibrium, or in other words is rounded by its own gravity. To look at Miranda is to see one of the most rugged landscapes in the Solar System, a world of extreme geological features carved into its icy surface, and it's here that we find the highest cliff face. Verona Rupes is reckoned to be 20 kilometres in height, and it's been estimated that in Miranda's low gravity it would take 12 minutes to fall from the top to the bottom of it.

The other four large moons around Uranus are less visually striking. All are thought to be constructed around a rocky core, surrounded by a mantle of ice, but these worlds are far from simple frozen lumps of rock: they all show signs of having been geologically active at some stage in their distant past – dark worlds in the distant Solar System that have surfaces shaped by volcanism. Ariel has a bright and seemingly young surface, paved over with recent and extensive flows of icy material in between the fault lines on its surface. Titania is also covered with huge geological faults and canyons, suggesting this too was a tectonically active world at some time in its history. With darker and potentially older-looking features, Umbriel or Oberon seem to be less active, or certainly less active in recent times.

Voyager 2 gave us our first chance to explore and understand the rings and moons of Uranus, but it was as the spacecraft widened its gaze that we were able to see in detail how this whole system is set in the most bizarre of orientations.

From our distant observations we had long thought that Uranus spun on its side, but to see the whole Uranian system up close for the first time made it clear that something very dramatic had happened in Uranus's past to set it apart from every other planet. Uranus spins on its side with an axial tilt of 97.77 degrees, resulting in one of its poles facing towards the Sun at any one time; the rings and moons are all oriented along this plane, too, so the whole system orbits the Sun like a giant heavenly bull's eye. We don't know for certain why Uranus is tipped over on its side, but it seems likely that at some point in the distant past it was hit by another planet, at least as large as the Earth, which quite literally knocked the planet over. Backed up by modern computer simulations, it is entirely feasible that such an impact would cause the rest of the system to follow, ending up with a planet, its moons and rings corkscrewing around the Sun.

The origin of Uranus still remains for the most part unknown, and the exact reason for this eccentric orientation remains one of the great mysteries of the Solar System. Voyager 2 left us with as many questions as answers, and when its encounter with Uranus officially came to an end with the firing of its thrusters on 25 February 1986, we left the planet and its secrets behind. Thirty years later, we still have no plans to return.

Our first and only close encounter with Uranus may have lasted for just a few hours, but it returned thousands of precious images and data. We had already gone further than ever before, and as Voyager left Uranus behind, it had ahead of it a lonely 1.6 billion kilometres before it would encounter another world.

Top and middle: The ice moon Miranda is the smallest of the Uranian satellites and one of the smallest studied objects in the Solar System, with a rugged, icy landscape.

Bottom: The five largest moons of Uranus. From left to right (in order of increasing distance from Uranus): Miranda, Ariel, Umbriel, Titania and Oberon.

Opposite: The five high-altitude clouds in Uranus's atmosphere visible here in orange on the right edge are almost as large as continents on Earth.

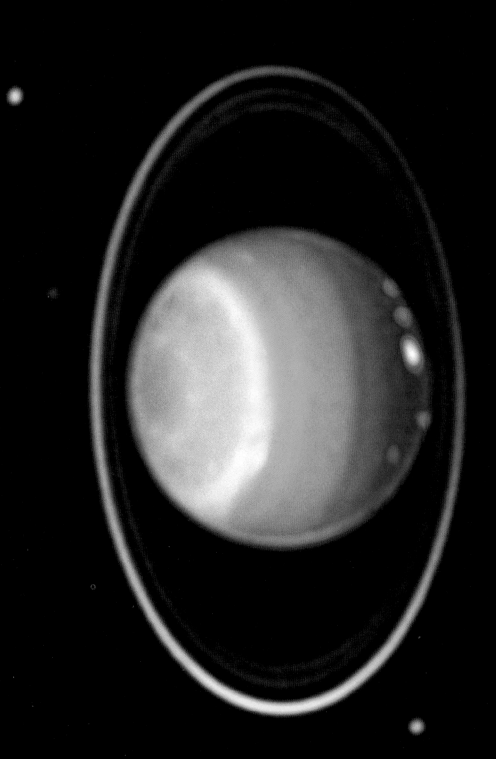

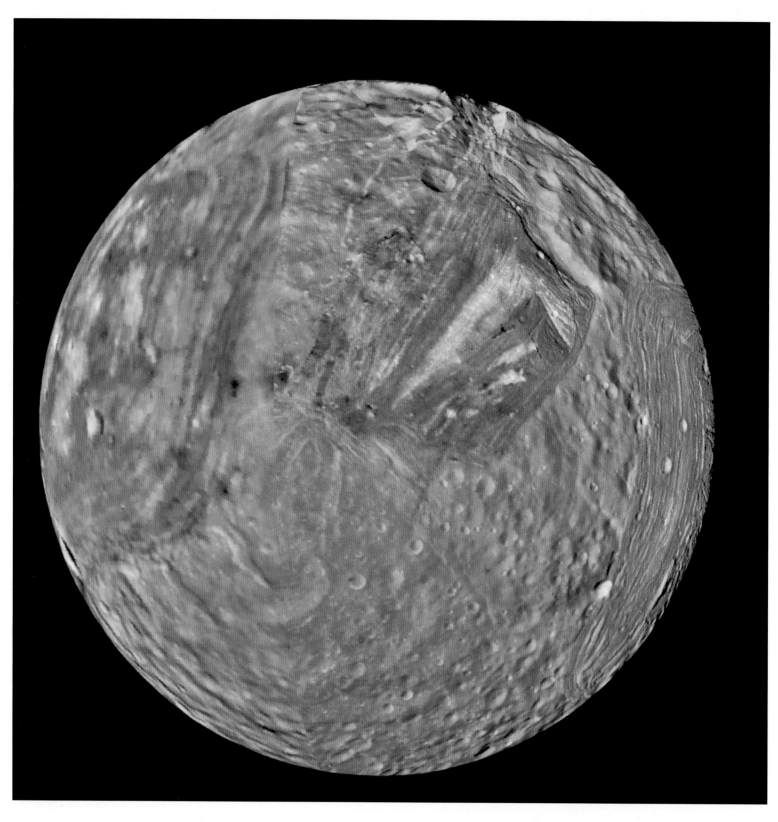

'Miranda was a world that had been broken up
and put back together. We saw huge cracks, cliffs,
impact craters, rifts – a real jumble of geology.'
Fran Bagenal, New Horizons mission

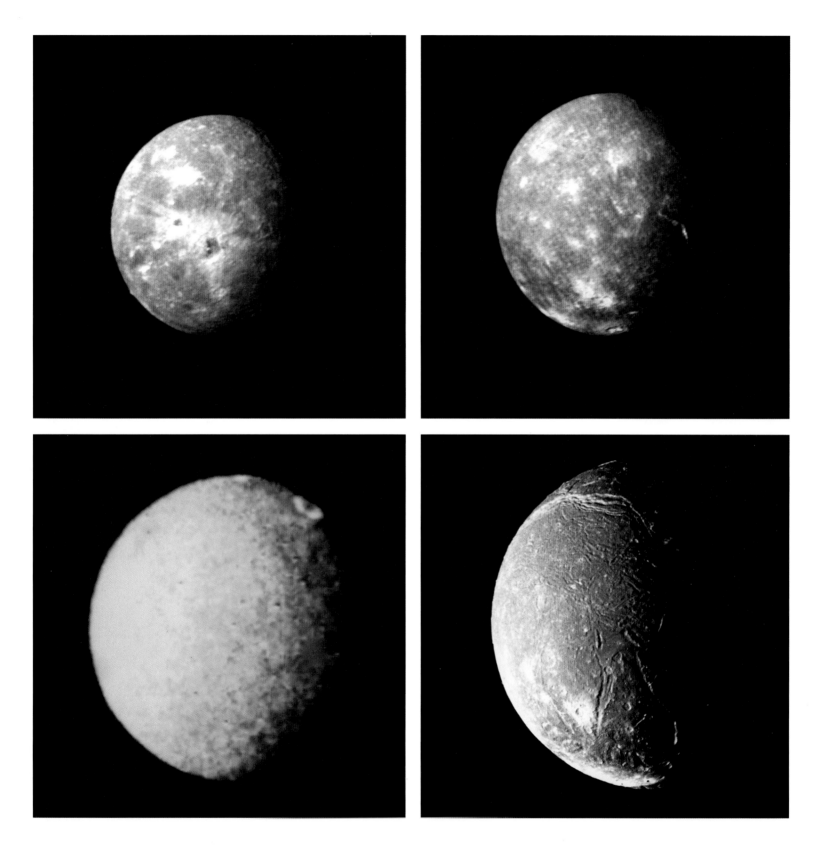

Top left: Oberon, the second largest moon of Uranus, was discovered in 1787 but little was known about it until Voyager 2's flyby.

Top right: Uranus's largest moon, Titania, was revealed by Voyager 2 to be geologically active, with a visible system of fault valleys.

Bottom left: This coloured image of Umbriel, the darkest of Uranus's moons, shows the impact craters covering its surface.

Bottom right: First spotted in 1851 by William Lassell, Ariel has the brightest surface of Uranus's moons, and possibly the youngest.

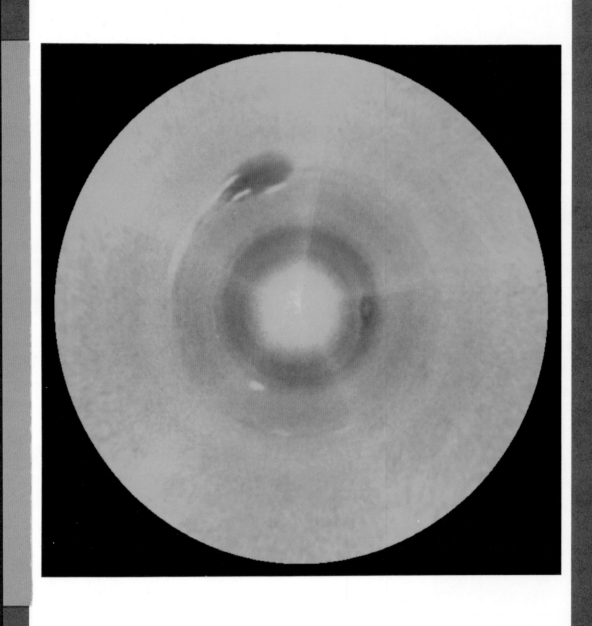

EXPLORING THE FROZEN REACHES

It would take three more years before Voyager 2 would reveal the first images of Neptune, the last planetary stop on its grand tour of the Solar System.

From a distance of over 57 million kilometres, our first images of this most distant of planets began to slowly come into view in early June 1989. With the near-featureless orb of its sibling Uranus still lingering in our thoughts, the planet emerged from the darkness, a tiny sphere with a thick blue hazy atmosphere, and it was immediately apparent that this was the more animated of the two ice giants. Taking 246 minutes for signals from the spacecraft to reach the Earth, early images sent back to us across the vast distances revealed enormous bands of high-altitude clouds floating above the blue haze. This was no inert, frozen world – it was a planet of extremes.

Until Voyager arrived in the Neptunian system, this was the least understood of all the planets. Invisible to the naked eye, Neptune was a world hidden from view. Even through the most powerful of telescopes it appeared as no more than a blurry ball of blue and green. The fourth-largest planet in the Solar System by volume and the third-largest by mass, from afar we knew nothing of its atmosphere or its structure and had only faintly observed two of the moons in its orbit – Triton and Nereid. This was a planet that had remained very firmly in the darkness, a distant world that had clung on tightly to its secrets for over 140 years.

Even the discovery of Neptune played out in the darkness. The last of the planets to be discovered, it is the only one whose existence we predicted before it was ever observed by the human eye. For centuries we stared at it without knowing what we were looking at. Almost certainly, Galileo was the first human to peer at Neptune directly, during a series of observations at the end of 1612 and beginning of 1613. The data he plotted with his newly invented telescope led him to believe this was a fixed star in the night sky, not a planet. Over the next 200 years many astronomers must have fixed their gaze on this dim star without realising the significance of its bluish light, but the first hint of its true nature would appear not in an eyepiece but in a set of astronomical tables.

In 1821 the French astronomer Alexis Bouvard published *Tables astronomiques*, a set of tables predicting the orbits of the three outer planets – Jupiter, Saturn and Uranus – based on Newton's Laws of Motion and Universal Gravitation. Over the next few years Bouvard's theoretical work would be tested against the movements in the night sky, and his painstaking calculations were rewarded with the confirmation that actual observations of the movements of Jupiter and Saturn adhered precisely to his mathematical predictions. However, Uranus refused to behave quite so compliantly. The movement of the then farthest planet deviated substantially from Bouvard's predicted path, and this led him to make a suggestion: either Newton was wrong, or perhaps there was an as yet undiscovered planet lurking in the darkness, distorting the passage of Uranus through its gravitational interaction. A flurry of astronomers sped to find the hidden planet – the ultimate prize for any stargazer – as an acrimonious race kicked off between the British and French to predict the planet's location.

On the British side of the Channel, the Cornish astronomer and mathematician John Couch Adams became obsessed with Bouvard's hypothesis and was convinced he could use nothing more than pen and paper to determine the size, position and orbit of this hidden world. By September 1845 Adams had at least partially completed

Opposite: Jupiter's south pole; the latitude lines in concentric circles are clearer in this colour composite, punctuated by swirling storm clouds.

Top: When Uranus refused to conform to Alexis Bouvard's predicted path he dared to suggest Newton's theory was wrong.

Bottom: Neptune's Dark Spot coming into view; like Jupiter's Great Red Spot, this is a large storm system. It was identified by Voyager 2.

his calculations, but unknown to him, almost simultaneously, the French mathematician Urbain Le Verrier in Paris was completing his own complex computations of Uranus's path and the elusive eighth planet.

By the summer of 1846 Le Verrier had played his hand and announced to the Académie des sciences on 31 August that he had calculated a predicted location for the missing planet. Revealing his work just two days before, Adams secretly sent his own prediction in a letter to the Royal Greenwich Observatory for verification. In what must be one of the most close-run races in scientific history, Le Verrier engaged the Berlin Observatory in the search for his planet, and at 00.15 on the night of 24 September 1846, Johann Galle, with the help of his assistant Heinrich d'Arrest, looked through his refracting telescope and became the first person to knowingly gaze at the eighth and furthest planet in the Solar System. A planet that had taken decades to locate was found in less than an hour and within 1 degree of the position Le Verrier had calculated. Even today the exact credit for the discovery of Neptune is still disputed, but in the winner-takes-all world of astronomical discovery it's Le Verrier whom history records as the discoverer of Neptune – the man who discovered a planet 'with the point of his pen'.

It took another 143 years before we would visit Le Verrier's planet and witness a dynamic, violent and volatile world. As Voyager drew closer it began to return

Right: Voyager 2 captured Neptune's Great Dark Spot, which was invaluable, as years later, this storm system would disappear from view.

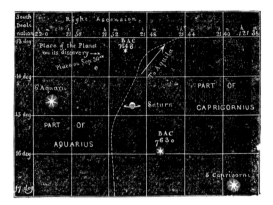

'After a few years the Spot dissipated, but then another one formed, so Neptune doesn't have this giant permanent spot like Jupiter does, it tends to develop big dark storms.'
David Grinspoon, astrobiologist

a precious set of images that remain to date our only close-range record of this planetary system. At first flying past Nereid, the third-largest of Neptune's moons, Voyager captured our only image of this distant world travelling on its massively eccentric orbit around the planet, but it was the images of Neptune itself that became increasingly astounding. Painted across the surface of the planet, Voyager revealed banks of methane clouds that were being whipped around by high-altitude winds at speeds of over 2,000 km/h. At least five times stronger than the most powerful winds ever measured on Earth, these were the highest wind speeds we'd witnessed anywhere in the Solar System.

It wasn't just violent winds that we saw through Voyager's eyes. Great hurricane-like storms could be seen moving across its surface, the largest being an Earth-sized storm system, similar to the Great Red Spot on Jupiter, which became known as the Great Dark Spot. Voyager took multiple images of this vast storm system as it sped past the planet, skimming the cloud tops of the northern pole from a distance of just 4,400 kilometres as it reached its nearest approach on 25 August 1989. No one had expected such massive storms to be brewing so far from the Sun, but just five years later, with the Hubble Telescope up and running, we peered back towards Neptune and found that the Great Dark Spot had disappeared. Unlike the giant storms on Jupiter, which we have witnessed playing out for centuries, it seems that

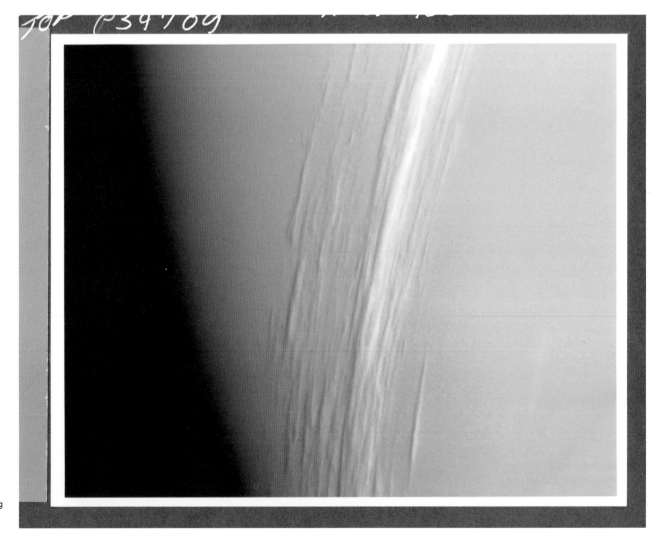

Right: Cloud streaks encircling Neptune, ranging in width from 50 to 200 km.

'Neptune is so aesthetic, so beautiful. We can't help thinking this is a bit like Earth, another blue planet. But it is blue for a very different reason; it has a methane atmosphere that absorbs the red part of the sunlight.'
Fran Bagenal, New Horizons mission

the weather on Neptune is far more transient, with the largest events coming and going within the space of a few years.

The intense weather systems that we witnessed on Neptune remain a tantalising puzzle. The planet is on average 50 per cent further from the Sun than Uranus and receives just 40 per cent of the sunlight that hits its sister world, and the surface temperatures of both planets are equally intensely cold. In fact the temperature of Neptune's outer atmosphere is one of the coldest places in the Solar System, with temperatures at its cloud tops measuring around minus 220 degrees Celsius.

Orbiting this far out in the far frozen reaches of the Solar System explains the blandness of Uranus's weather, but why would Neptune, even further from the Sun, be so active? The answer, we think, lies in a mysterious and powerful source of heat that we have detected coming from deep within the planet.

All of the planets, including Uranus and Neptune, have some residual heat locked deep inside them, and much of this is the thermal energy that remains from the endless collisions of their creation, as well as the heat released by the radioactive decay of certain elements in the core. However, this far out Uranus has rapidly lost this heat and today radiates just 1.1 times the energy it receives from the Sun, so that the heat radiating from its core is too little to drive any kind of dynamic weather system through its atmosphere. Neptune, it seems, has a far warmer heart, and Voyager was able to reveal that the most distant planet radiates 2.61 times as much energy as it receives from the Sun.

It's these unexpectedly high temperatures that help to explain the violent storms we see in its atmosphere. As the heat makes its way from the core of the planet and out into space, it churns up the entire atmosphere, creating winds of extreme ferocity – and once they start there is little to stop them. Neptune, like Uranus, is an ice giant, a planet composed of a rocky core and a mantle made up of water, ammonia and methane ices, surrounded by a cold gaseous atmosphere. With no solid surface on the planet there are no mountains or continents to break up the flow of the atmospheric gases, so this means the winds can whip around the planet, picking up speed until they are quite literally supersonic.

All of this helps to give us a reason for Neptune's extreme weather but not an explanation. Voyager 2 was unable to reveal why Neptune and Uranus are so different, and we still don't know what it is about these planets' formation that left Neptune with far more residual heat, making it the far more active of the two.

Neptune's active atmosphere was not the only surprise Voyager returned to us with the pictures from its two-camera system. In vibrant and vivid detail the images also revealed the Solar System to be home to not one but two blue planets. Neptune is a bright blue world, but with no liquid water on its surface we know that the origin of this colour cannot be the same as here on Earth. This deep cobalt colour must be produced by something else in Neptune's atmosphere, and Voyager provided the answer. Using its on-board spectrometer Voyager was able to measure for the first time the composition of Neptune's atmosphere and found it to be comprised of 80 per cent hydrogen, 18.5 per cent helium and 1.5 per cent methane. Methane was by far the smallest component, yet this is the key to understanding Neptune's blue hue.

ANATOMY OF NEPTUNE

49,244 km

rocky, icy core

inner mantle (methane, ammonia and water)

outer mantle (mainly liquid hydrogen)

atmosphere (hydrogen, helium and methane)

Top left and right: Neptune's Great Dark Spot and the rotations of the clouds around it.

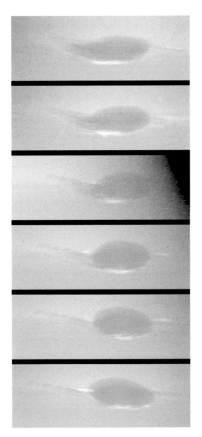

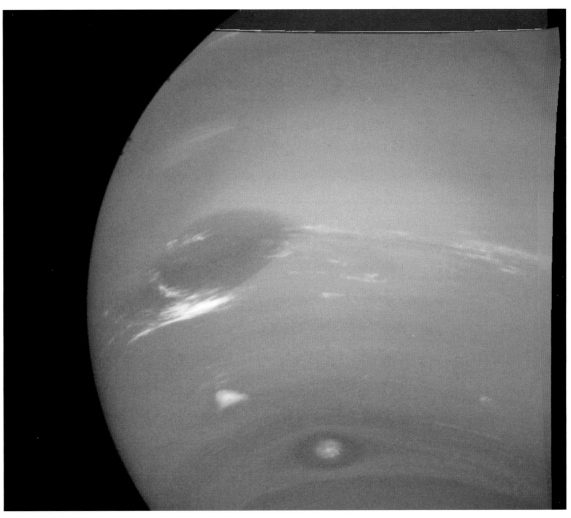

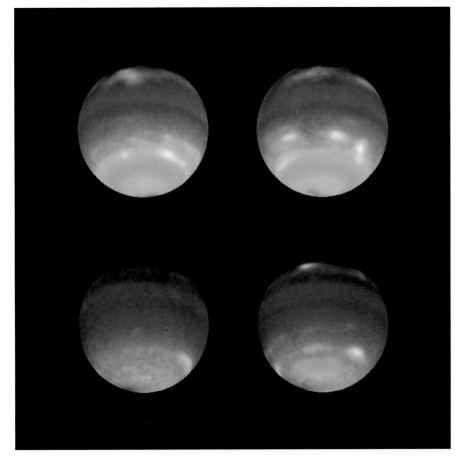

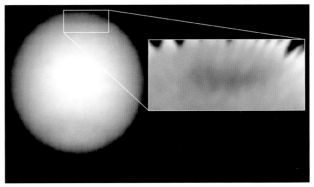

Left: Time-lapse images of Neptune's rotation, which allow scientists a close view of the ebb and flow of the planet's weather systems.

Above: The Great Dark Spot as seen through the Hubble Space Telescope, with a close-up of the high-pressure system.

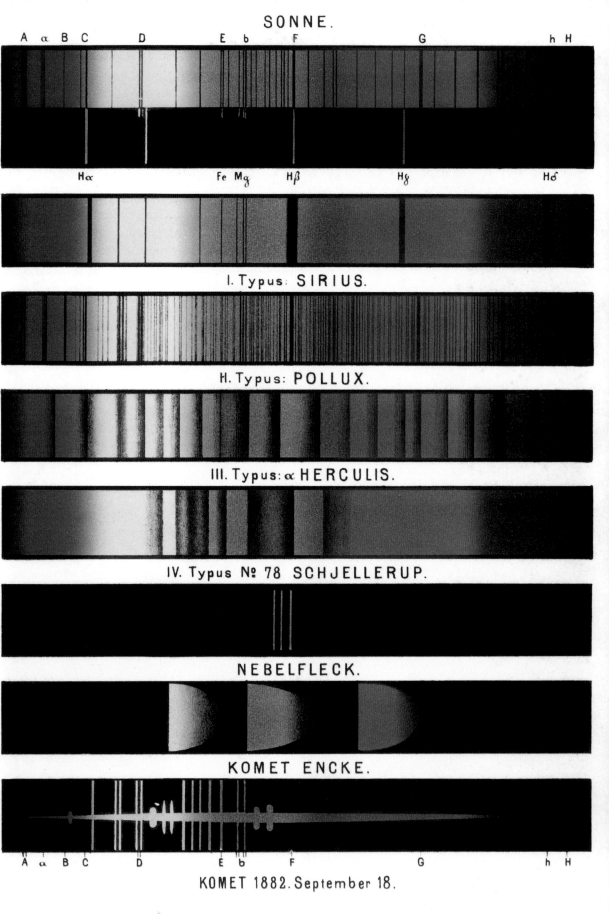

Left: The light of stars contains all the colours of the rainbow, but how light is absorbed or reflected depends on elements present in the atmosphere. This illustration compares the light absorption spectrum of our Sun (top) with other stars and comets in the galaxy.

Opposite: This large freshwater lake produces a natural source of methane hydrate, which gives an insight into the landscape of methane-packed Neptune

'We have a basic understanding of the colour of Uranus and Neptune. These ice giants, in addition to hydrogen and helium in their atmospheres, have a healthy amount of methane in their upper atmospheres, which gives the blue colour.'
David Grinspoon, astrobiologist

The light of the Sun contains all the colours of the rainbow, but as it hits the surface of a planet it interacts in different ways depending on the molecules it encounters. In the case of the Earth, the water that covers our planet absorbs all of the red light in sunlight, so as the Sun strikes the surface only the blue is reflected back. It's the same with all the plants on our planet – the chlorophyll absorbs the blue and red light so that only green is reflected back. On Neptune there is no water to turn the planet blue; instead it's the methane that absorbs the red light of the dim and distant sun, reflecting the blue light back out into the darkness, but things aren't quite as simple as they at first appear. Uranus is also a blue world, but just that bit paler and greener in shade compared with its more distant sibling, and yet Uranus also has methane in its atmosphere – in fact it has slightly more – so Uranus not Neptune should be the darker shade of blue. Why is this? Well, quite wonderfully, we still don't know. Voyager looked for other constituents in Neptune's atmosphere that could account for the deeper blue but it found nothing that could explain the discrepancy, so for now the many differences between these two worlds remain a mystery.

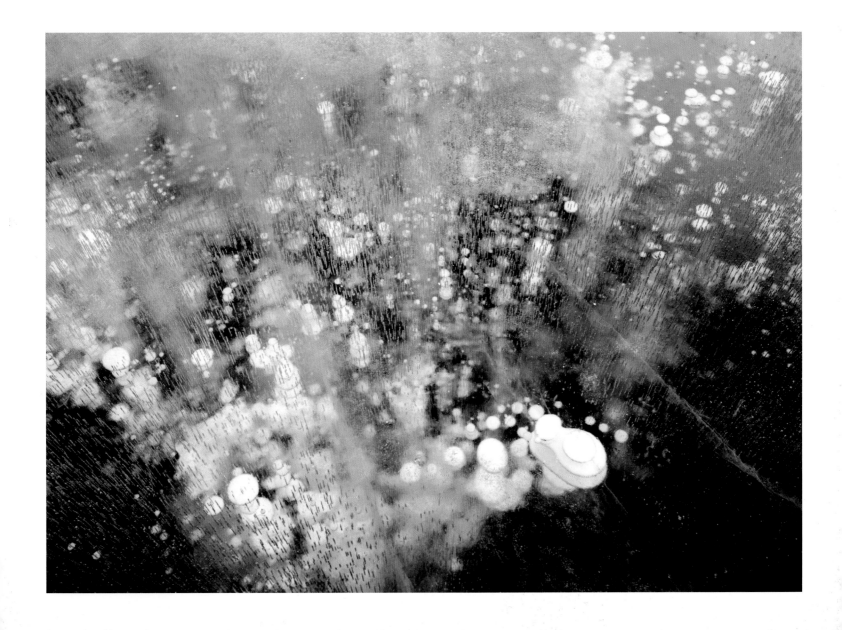

NEPTUNE'S MOONS

After over 12 years of travelling, Voyager 2 completed its grand tour of the Solar System, a tiny explorer that had set out from Earth and travelled across 7 billion kilometres of space on a single extraordinary journey to explore all of the outer worlds of the Solar System – Jupiter, Saturn, Uranus and Neptune. But before it began its long, lonely journey into the darkness of interstellar space there was one more brief rendezvous left on its itinerary.

While in the Neptunian system Voyager had made endless discoveries, including the detection of a faint ring system and the confirmation of six new moons in orbit, but in those final hours of close contact, as it flew across the north pole of the planet, the mission scientists back at JPL had instructed the craft to make one last intrepid manoeuvre: a close flyby of Neptune's giant moon, Triton.

Just five hours after nearly touching the cloud tops of Neptune, Voyager travelled within 40,000 kilometres of the Neptunian moon and returned the first pictures of this strange and distant world.

The British astronomer William Lassell first spotted Triton on 10 October 1846, just 17 days after the discovery of Neptune. As astronomers clamoured to observe the newly identified planet, it was Lassell with his homemade telescope who noticed the satellite first. Although not officially named until many years later, Triton would remain the only known satellite of Neptune for over 100 years, until the discovery of Nereid by Gerard Kuiper (more of him later) in 1949.

As Voyager returned the first images of this distant moon it immediately became apparent that, just like Neptune, Triton was far more active than we had ever imagined. Here in the furthest reaches of space the seventh-largest moon in

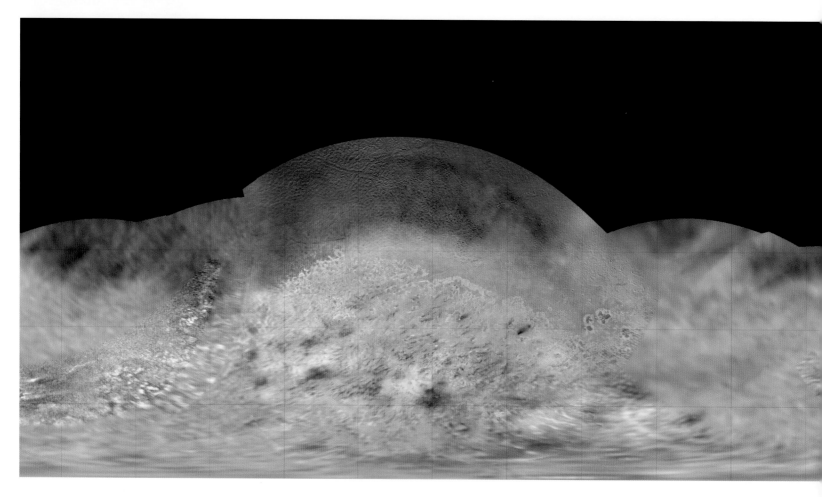

Top: This global colour mosaic of Triton taken by Voyager gave scientists a fascinating glimpse of the moon's surface.

Bottom right: Polar projection of Triton's southern hemisphere showing the polar cap and equatorial fringe.

'The final place that Voyager visited was Neptune's moon Triton, and it really was a spectacular finale.'
David Grinspoon, astrobiologist

Opposite top: The largest of Neptune's 13 moons, Triton is the only large moon in our Solar System with a retrograde orbit – it orbits in the opposite direction of its planet's rotation.

Opposite bottom: Data obtained from Voyager enabled scientists to produce this global colour map of Triton.

the Solar System was far from a frozen, inactive rock, it was a geologically active world with features unlike anything we had imagined. Able to image just 40 per cent of its surface, due to the fact that most of the northern hemisphere was in darkness, Voyager revealed Triton to have a sparsely cratered exterior with a strange network of ridges and valleys covering its entire surface – including in the western hemisphere a region dubbed the 'cantaloupe' terrain because of its striking similarity to the skin of a melon.

All of this evidence pointed to the fact that the surface of Triton was 'young', not battered and ancient like that of our Moon – a surface that, even though it had been recorded by Voyager's instruments as one of the coldest places in the Solar System at minus 235 degrees Celsius, seemed to be being resurfaced by active volcanism occurring across the planet. But this was not volcanism as we understand it here on Earth; there is no molten rock beneath the surface of Triton. This is an ice world with a surface covered in a frozen mixture of nitrogen, water ice and carbon dioxide. Any volcano that exists here would be a cryovolcano, erupting water, ammonia and methane from below the surface – cryomagma that would immediately solidify in the extreme low temperatures at the surface.

And as Voyager peered closer at the surface of Triton, it captured a series of images of the southern polar region that appeared to provide evidence of such geological activity. Lurking amongst the flat volcanic plains and rift valleys was a set of distinct features that at once caught the eye: scattered across the surface were what appeared to be at least 50 dark plumes. Voyager was travelling so fast and so far away from the moon that it was unable to get close-ups of these intriguing plumes, but we are

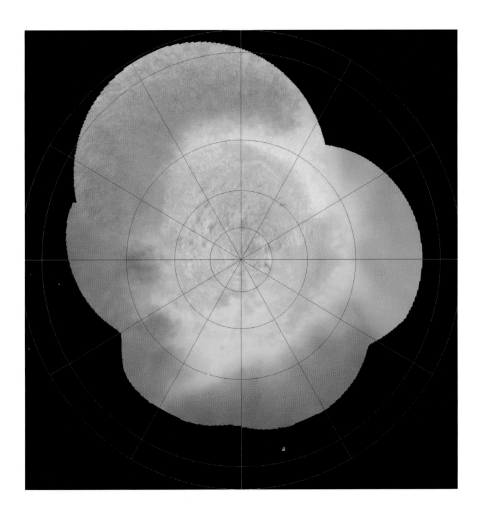

certain that they are the evidence of ice geysers, and that the dark streaks are dust deposits left on the surface by the eruption of these geysers. According to planetary scientist Noah Hammond, 'It was a big surprise to see active geysers going on on Triton, shooting 8 kilometres into the air, and we think these geysers might be forming as a result of nitrogen ice which is being vaporised by the Sun. Both of the geysers that Voyager 2 observed, which are currently blowing up, are concentrated on the warmest part of Triton, where the sunlight is striking the hardest. And as the sunlight pierces through the nitrogen ice it starts to turn that nitrogen ice back into a gas again, and that gas builds up and builds up until it has enough energy to break through the ice and lift dust all the way up into the atmosphere.'

Further analysis of the images suggested that at least four of these geysers were active at the time of Voyager's flypast, belching clouds of ice and dust 8 kilometres up into the thin Neptunian atmosphere before falling back down to the surface over 100 kilometres downwind. Perhaps even more astonishingly, we have been able to actually formulate a hypothesis for the mechanism that lies beneath them. The key to understanding why this wondrous geological activity occurs in the most unlikely of places seems to involve their location on the surface of Triton.

At least four geysers were active at the time of Voyager's flypast, belching clouds of ice and dust 8 kilometres up into the thin Neptunian atmosphere.

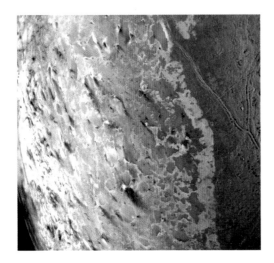

From the imaging data that Voyager 2 returned, it seems the geysers tend to erupt on the most sunlit parts of Triton's surface. From a distance of 4.5 billion kilometres the Sun's rays are faint and the amount of thermal energy landing on the surface is low, but combined with the specific chemistry on Triton's surface the effect of the Sun can still be dramatic.

What seems to be happening is that sunlight falling on the thin layer of nitrogen ice that covers Triton is penetrating the surface and heating up a layer of darker methane particles a metre below the surface. The methane acts as a subsurface greenhouse gas, trapping the heat below the surface and radiating that heat back up through the frozen nitrogen. Even though this mechanism creates a temperature difference of just 4 degrees Celsius, between the ambient surface temperature and that of the warmer interior this difference is enough to melt the frozen nitrogen and create pockets of gas. Sitting beneath the nitrogen ice, that gas is kept under pressure, building up until ultimately it punches through the nitrogen ice, creating the geysers of Triton and carrying the dark ice particles up to 8 kilometres into the sky.

Although the faint light of the Sun is just strong enough to power Triton's geysers, it is simply too weak to have driven the other geological processes that have created the network of valleys and crevasses we see all across the surface. It seems that something in Triton's past must have heated the moon enough to drastically alter its terrain, and a potential clue to the source of that ancient heat can be found in the oddity of Triton's orbit.

Unlike every other large moon in the Solar System, Triton orbits in the opposite direction to the spin of the planet. Travelling around Neptune in its tidally locked orbit, with the same side facing inwards at all times, Triton orbits in a clockwise direction while Neptune spins on its axis every 16.11 hours, anti-clockwise. Such dissonance between planet and satellite suggests that it is very unlikely that Triton and Neptune formed at the same time. A planet and moon system fashioned from an identical collapsing cloud of gas and dust at the same moment would tend to spin and orbit in the same direction as the spin of the initial dust cloud, just like the overwhelming majority of moons we see in the Solar System today. So the most likely explanation is that Triton joined Neptune much later, a visitor to the Neptunian system that, unlike Voyager, never left. Like Uranus, we don't know for certain the deep history of Triton, but all the evidence from its orbit and surface features point to the possibility that long ago Triton was not a moon at all. To understand where it might have come from we need to look beyond the Neptunian system and peer further out into the darkness.

It was only relatively recently that we confirmed the existence of a region of the Solar System now known as the Kuiper Belt, a disc of millions of objects that stretches out from the orbit of Neptune to at least 50 au from the Sun. Similar to the asteroid belt in structure, but far larger in scale, we had long speculated that such a region might exist, but it wasn't until the discovery of an icy object known as 15760 Albion, in 1992, that we were able to put together a body of proof. The discovery of this region was a turning point in our knowledge about our Solar System. According to Alan Stern, of the New Horizons mission: 'The Kuiper Belt was so revolutionary because as it's so far from the Sun, the temperatures out there are almost absolute zeros, so everything is so well preserved. It's an archaeological dig, if you will, into the early history of our Solar System, so it's a scientific wonderland, and you ... put all

Above: The south polar terrain of Triton showing dark plumes or wind streaks on the moon's surface.

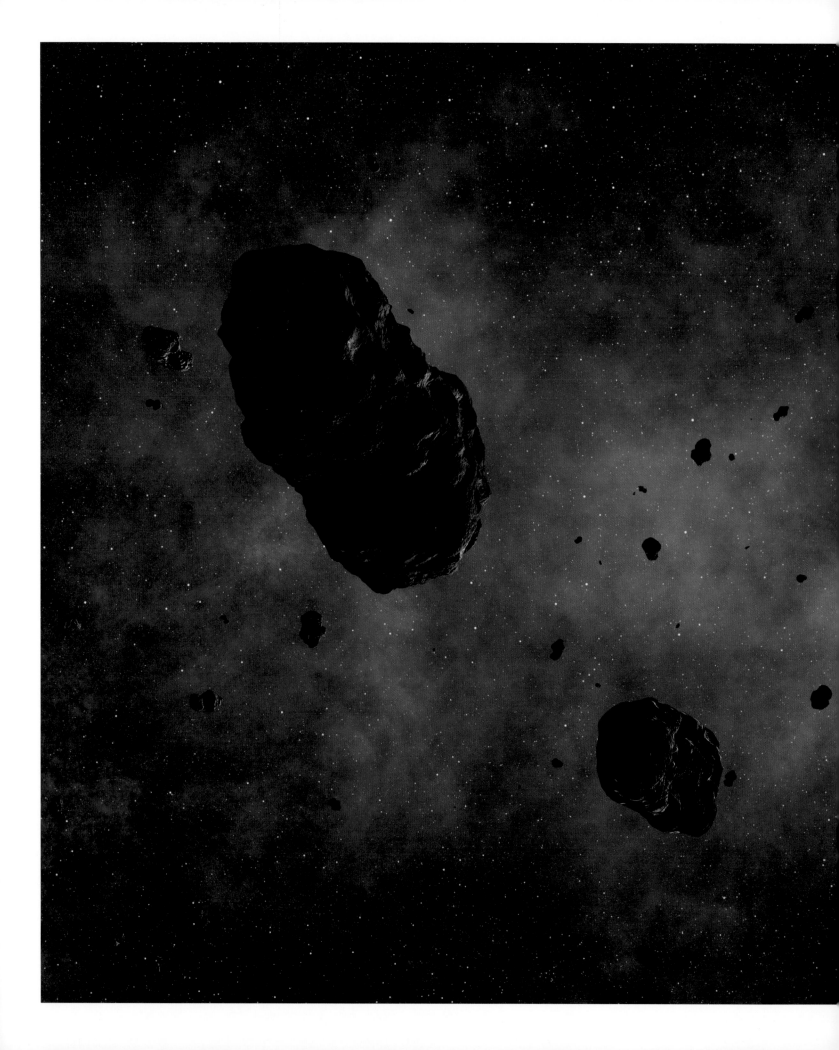

that together, redrawing the map and putting us in our place about who's the oddball now. It's not Pluto, it's the Earth or Jupiter that's the oddball, and then having this museum piece place in the Solar System, where everything is so well preserved, I can't think of a bigger discovery that we have made, that's reshaped our thinking so profoundly in all the decades of space exploration.'

Albion was the first trans-Neptunian object to be identified since the discovery of Pluto and its moon Charon, and this finding launched a flurry of further revelations that now put the number of identified objects in the Kuiper Belt region at around 2,400. We think there are at least 100,000 objects bigger than 100 kilometres wide in the Kuiper Belt, but we'll explore our recently expanded knowledge of this region later in the chapter. Just like the asteroid belt, this dispersed region of debris is composed from the remnants of our Solar System's formation, but whereas the asteroid belt is a disc of rocky fragments, we think the Kuiper Belt is made up of icy objects, with methane, water and ammonia the major ingredients, and it's that list of ingredients that leads us back to Triton and its strange history. We believe now that perhaps only Triton grew up out in the Kuiper Belt, amongst the millions of other frozen lumps of water, ammonia and methane circling the Sun, before something such as a collision disturbed its orbit and flung it inwards. Plucked from the Kuiper Belt, Triton would be captured by the gravitational pull of Neptune and forever trapped in its strange orbit around the blue planet.

Today Triton orbits in a nice regular circle around Neptune, a bit like the way that our moon orbits around the Earth, but to begin with this newly captured satellite would have almost certainly been in a chaotic orbit. Glancing past the planet, it would have started along a wide elliptical path around Neptune, bringing it from far out to close in on every revolution. This would have meant that as Triton orbited Neptune the gravitational pull would be constantly changing, stretching and squashing the moon, just as Jupiter does to Io today, with the result that the friction of these tidal forces would have heated up the interior of Triton, creating far more activity than the erupting geysers we see today. Through this process of tidal heating the ancient interior of ice would have become molten and volatile, exploding up through cracks and faults in the moon's crust, creating the ragged surface we see today. Over time, as the orbit adjusted into a more regular, circular path, the friction reduced, the heating diminished, and so what we are left with is an icy, frozen moon, circling the wrong way around its planet, the subtle scars of its early life written across its surface. All that remains today of that dramatic violent past are the geysers, powered by the faint light of the Sun, painting dark streaks across the now quiescent surface.

As Voyager 2 sped away from Triton and Neptune at over 60,000 km/h the planetary exploration phase of this extraordinary mission finally came to an end. Its exploration would continue beyond even the furthest reaches of our Solar System and would extend into the unknown depths of interstellar space, but its time as a planetary explorer was over. A tiny spacecraft, built with a thousand human hands, had shone a light into the darkness, revealing the secrets of worlds that had been hidden in the night for over 4 billion years. As yet we have not returned, but while Neptune and Uranus have fallen back into the distance, viewed only through the lenses of our most powerful telescopes, another even more distant world was out there waiting for our first arrival.

Opposite: Further exploration of the vast and mysterious Kuiper Belt might unlock some of the secrets of the origins of the Solar System.

SEARCH FOR PLANET X

On the night of 23 January 1930, a 23-year-old junior researcher at the Lowell Observatory in Flagstaff, Arizona, took an image of the night sky using a 13-inch astrograph, a telescope designed for the sole purpose of astrophotography. Clyde Tombaugh had been at the observatory for just over a year when he took the image, one of hundreds he had captured over previous months. A keen amateur astronomer, he had landed this first job after sending in a series of drawings he had made of Jupiter and Mars using telescopes he had built at his home. Every night at the Lowell, Tombaugh performed the same task, using the astrograph to photograph a precise section of sky that he would then image again a few nights later. The task was a little laborious but its purpose was profound, because hiding within the near-identical images Tombaugh was on the hunt for a new world, the enigmatic planet X.

By 1930 the search for the elusive ninth planet had been running for well over three decades. Since the turn of the twentieth century, precision observations of the orbits of Uranus and Neptune had suggested there was something else lurking in the darkness of the outer Solar System, another distant world far beyond Neptune. But hunting down such a faint object was like looking for a needle in the celestial haystack.

Leading the search for planet X over much of this time was the founder of the Flagstaff Observatory, Percival Lowell. Starting in 1906, Lowell had embarked on an obsessive pursuit of the ninth planet, but despite years of work he would not live long enough to see its discovery and he would never know how close he came to finding it himself. Unknown to Lowell, an image of planet X had been captured in two photographs he had taken just a year or so before his death in 1916, but the evidence was missed – the photographic plates were filed away and the search would go on without him for another 15 years.

The logic behind using identical pairs of photographs taken a few nights apart was simple: by examining the two images for any stars that appeared to have shifted position, they would be able to differentiate between the distant stars in the background and anything that was orbiting closer to Earth. Hundreds of pairs of images were taken at the Lowell Observatory for this purpose and analysed using a blink comparator, which helped to reveal any minute differences between the images, but it wasn't until the night of 29 January 1930 that Tombaugh would strike gold.

Comparing the two images he had taken a week apart, Tombaugh discovered what appeared to be a moving object on the two plates. Across those few days a dim spot had moved no more than an inch, but that movement signified something profound. This was an object much closer to the Earth than the distant stars in the background, an object that had moved as our perspective had changed with the Earth's own orbit, and over the course of that week this object appeared to have shifted against the distant background. Tombaugh had found planet X, which would be named Pluto, after the suggestion of Venetia Burney, an 11-year-old English schoolgirl from Oxford with an interest in classical mythology.

1930. MARCH. 2ₚ. 4ₕ 56ᵐ

1930. MARCH. 5ₚ. 3ₕ. 4ₕ.

Above: Early images of Pluto, taken in March 1930.

Top: Clyde Tombaugh's dedication unravelled the secrets of Planet X, which he spotted in the night sky in 1930.

Opposite top: The two telescope images that astronomer James Christy used to identify Pluto's moon Charon in June 1978.

Opposite bottom: Detailed views of almost all of the surface of Pluto, constructed from images taken by the Hubble Space Telescope.

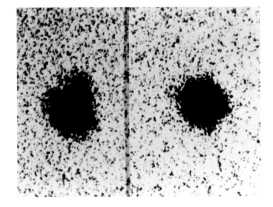

'The discovery of the Kuiper Belt in Pluto's context in the 1990s was probably the single most important discovery about our Solar System, of the space age ... it redrew the map completely.'
David Grinspoon, astrobiologist

Its discovery made headline news around the world, propelling the unassuming Clyde Tombaugh into the most exclusive of astronomical clubs. Along with William Herschel and Urbain Le Verrier, Tombaugh would go down in history as one of only three named humans to have found a planet – or at least that's what he and the rest of the world thought. He died in 1997, just before his 91st birthday, unaware of the controversy that would unpick the status of his discovery just a few years later.

Tombaugh would live out his whole life intimately linked to the speck of dust he'd spotted on that photographic plate, a distant rock that would remain for decades more vivid in our imaginations than in our sights. Even with our most powerful of telescopes straining to see into the dark, the best image that our twentieth-century technology could provide us with was nothing more than a fuzzy blur, just a handful of pixels. The Hubble Telescope, freed from the distortions of our atmosphere, could peer further, but it provided us with little more to see, just the most frustrating of glimpses of an orb hanging in the darkness with all of its features hidden from view. Well into the twenty-first century Pluto remained out of reach, a world too far for us to understand or explore, but as with all of our greatest explorations, that blank spot would be removed in an instant.

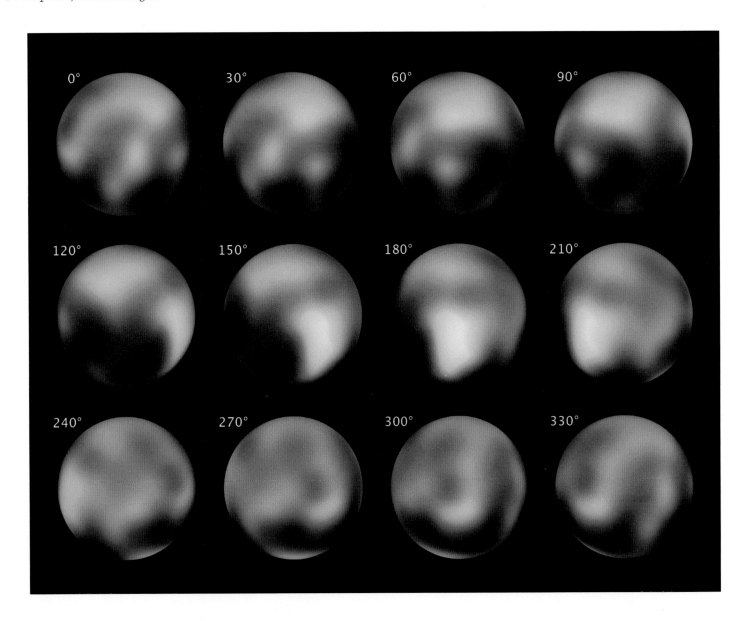

SEARCHING NEW HORIZONS

'New Horizons had to work perfectly. It was one spacecraft travelling by itself billions of miles over nine and a half years, with no backup. Pluto was one shot. There was no making a U-turn and coming back for a redo.'

Alan Stern, New Horizons mission

Taking off from Cape Canaveral Air Force base on the back of an Atlas V rocket, the New Horizons space probe began its epic journey across the Solar System towards Pluto in January 2006. Just nine years after the death of Clyde Tombaugh, this tiny interplanetary probe no bigger than a grand piano made room in its payload for 30 grams of his ashes, a tribute to the man who had begun our journey to Pluto 76 years before. For Alan Stern, Principal Investigator of the New Horizons mission, this move fulfilled a promise made to Tombaugh; 'Shortly before New Horizons became a project, while we were still trying to get funding to fly a spacecraft to explore Pluto, [Clyde] died, in January of 1997, and prior to that he had asked, if a mission ever did get launched, if there was a way it could be considered if some of his ashes could be sent on the journey. And I had heard about that back in the 90s, and when it came close to launching New Horizons in the summer of 2005, I reached out to his family to see if that's something they wanted and it was still possible and it was, and we found a way to do it, and we were very proud to be able to make it happen. Not just for Clyde Tombaugh and his family, but even more so for the people of Earth, because it makes a very human connection between the discoverer of this world and those who were the explorers of it on the New Horizons team.'

Leaving the Earth with an escape velocity of 16.26 km/s, New Horizons was the fastest probe ever launched from our planet. Reaching Jupiter in just over a year, it progressed on its way with a gravity-assisted kick from the gas giant, sending it hurtling across space. With all non-essential systems shut down to conserve energy, this little probe would sleep through most of its epic eight-year voyage in the dark. After 20 years of planning, building and dreaming, the New Horizons team could only wait and hope for its safe passage as it made its way towards the last unexplored world in our Solar System. But while the voyage itself proceeded without a glitch, back on Earth the status of the mission was about to take the most unexpected of turns.

Starting with the discovery of the rocky object 15760 Albion back in 1992, the description of Pluto as a planet, orbiting alone in the farthest reaches of the Solar System, would come under increasing scrutiny. Pluto, it seemed, was not so lonely after all. This was a region of space filled with an array of orbiting objects, and as we began to peer into the Kuiper Belt we discovered they were far from just fragments; this was a region of space filled with Pluto-like worlds:

> 50000 Quaoar, an object half the size of Pluto, discovered at the Palomar Observatory in June 2002.
> 90377 Sedna, a large minor planet discovered in November 2003 on an 11,400-year orbit around the Sun.
> 136198 Haumea, a huge pebble-shaped world, with two tiny moons in orbit, discovered in December 2004.
> 136472 Makemake, a dwarf planet perhaps two-thirds as big as Pluto, discovered in 2005.

Then in January 2005 we spotted 136199 Eris, a world almost the same size as Pluto itself, if not bigger. These discoveries and the hundreds we've made since forced the International Astronomical Union, the ruling body on Earth for all things

Opposite and above: The New Horizons space probe, which ventured out to Pluto and the Kuiper Belt for the first ever mission to this part of space.

Below: This image gave the green light to scientists as a clear path for New Horizons to reach Pluto (right).

Below bottom: The relative sizes of Pluto's moons.

THE PLANETS

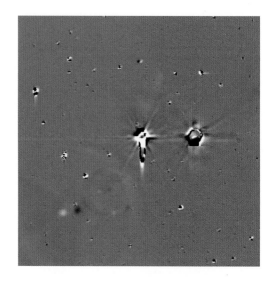

DWARF PLANETS

Earth's Moon | Eris | Pluto | Haumea | Makemake | Ceres

astronomical, to formally reconsider the definition of a planet. In October 2006 they published a set of three criteria that any object in our Solar System would have to fulfil if it wanted to stake a claim of planethood.

1. It has to orbit the Sun;
2. It has to be massive enough for it to form a sphere under its own gravity;
3. It has to be massive enough to clear its own orbit of any other objects, dust or debris.

Pluto certainly ticks the first box, completing an orbit of the Sun every 248 years. We also know, even from our most distant observations, that Pluto is nearly perfectly spherical, and so the second criterion is also a clear tick for Pluto. But when it comes to the third requirement, Pluto is not so lucky; it is simply not massive enough to blow away all the bits of dust and debris in its orbit, and for that reason it fails to make the cut as a planet. Since September 2006, Pluto has been officially designated a dwarf or minor planet.

A sphere orbiting the Sun, yes, but a planet, no. The third criterion is ultimately the one that necessarily separates the chosen eight – Mercury, Venus, Earth, Mars, Jupiter, Saturn, Uranus and Neptune – from the rest of the globes that orbit the Sun. Because if Pluto was a planet, so too would be Sedna, Eris, Makemake and all the other tens – or potentially even hundreds – of smaller objects that orbit in the distant Kuiper Belt, billions of kilometres from the Sun.

So for now at least we have a solar system filled with eight planets but many worlds – hundreds of moons and dwarf planets, many that remain unknown and out there waiting to be discovered. As for Pluto, despite the best efforts of complaint and protest around the world, by the time New Horizons entered the final leg of its epic voyage it had changed from a mission of planetary exploration to the first ever mission to a dwarf planet in the Kuiper Belt.

Thankfully our Solar System pays little heed to the fanciful definitions of humans from the International Astronomical Union, and while downgraded by name, the world that awaited us was anything but a disappointment.

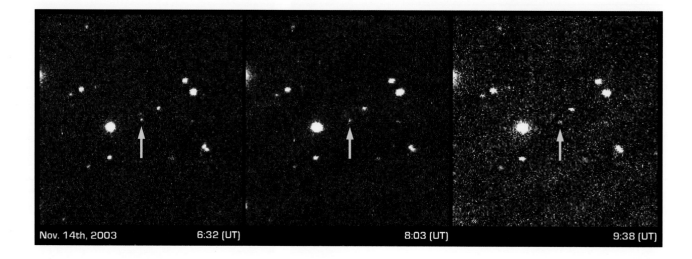

Nov. 14th, 2003   6:32 (UT)   8:03 (UT)   9:38 (UT)

Above: The discovery of a faint distant object in November 2003, which would later be identified as the minor planet Sedna.

Above: The second largest dwarf planet Eris (after Pluto) and its moon Dysnomia.

Right: The dwarf planet Quaoar, which is roughly half the size of Pluto, and its moon Weywot.

Above: Pluto in all its dwarf
planet majesty, as viewed
by the New Horizons probe.

'New Horizons redrew the map and then put us in our place; it told us that the giant planets and the Earth-like terrestrial planets are in the minority. It's the little dwarf planets that dominate the population of our Solar System.'
Alan Stern, astrobiologist

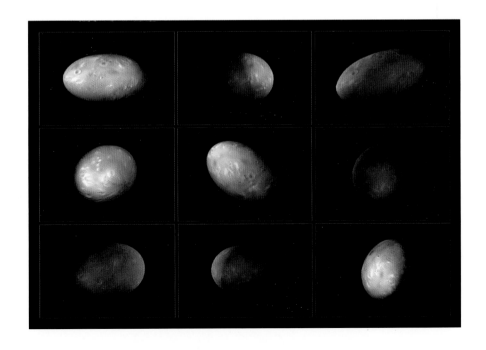

Above and top: The largest of Pluto's five moons, Charon. The image below shows the moon set against a star field, lit by the reflected light of Pluto.

Right: Pluto's moon Nix, shown orbiting Pluto-Charon and changing unpredictably as it does so.

Top: This image shows patterns of convection in an experiment on heat dissipation. Similar patterns were observed on Pluto's plains, indicating geological activity.

Bottom: The New Horizons flyby returned the closest ever images of Pluto's surface, revealing cratered, mountainous and glacial terrains.

After nearly ten years alone in the icy darkness of space, New Horizons awoke from its state of hibernation and the pixels that Hubble had seen all those years before were at last replaced by our first glimpse of the most distant world we have ever visited. In July 2015 New Horizons began to beam back images from its close encounter, to reveal Pluto not as a frigid, featureless, frozen rock but something far more beguiling. Planet or not, Pluto was beautiful! The images revealed a complexity of character stretching across its freezing minus 230 degrees Celsius surface that contained all the features of a dynamic, active world. The blemishes we'd gazed at from a distance were no longer just smudges but high-definition images revealing an extraordinary range of geological form and structure.

'New Horizons first spotted Pluto shortly after launch, just a dot in the distance. And year by year that dot grew a little brighter, but even after nine years of flight it was only 2 pixels across, and we were only ten weeks from arrival when the blurry images were as good as the Hubble had been able to do all that time. But then they got a little better and a little better and a little better and suddenly overnight, as we flew past, it went from kind of a naked-eye vision of what the moon looks like, where you can see the general patterns of bright and darkness, to all of a sudden this phenomenally complex world, with mountain ranges and gorges and glaciers, and an atmosphere glowing up to half a million feet in the sky, and polar caps, evidence for a liquid ocean on the inside, five spectacular moons and so much more. It was almost overwhelming in the first days and weeks after, to see what an amazing place Pluto was. And I took to saying that the Solar System had saved the best for last.'
Alan Stern, New Horizons mission

This was a landscape built from frozen nitrogen, with soaring mountains and complex ridged terrain with a look of snakeskin, but one feature more than any other leapt from the images New Horizons fed back to us: running across the equatorial region of the planet we could see a giant, heart-shaped plane of frozen nitrogen, methane and carbon monoxide that stretched for a million square kilometres, a region that has been named Tombaugh Regio, or to give it its more popular name, Pluto's Heart.

The western lobe of the heart is called Sputnik Planitia, and at its eastern edge lies a range of mountains made of pure frozen water ice that rises miles above the frozen plane. But it wasn't just the soaring mountains and romantic form of this region that caught the eye: there was something else about Sputnik Planitia, a strange detail that set it apart from the rest of the dwarf planet. Everywhere we looked across the surface of Pluto, we saw a world covered in craters – the scars of impacts that, just like on the surface of our moon, have taken place over billions of years. Except on Sputnik Planitia; this icy plain appeared unblemished, an utterly smooth surface, untouched for mile upon mile without a single crater to be seen. And the closer we looked at this region the stranger it appeared, because as we studied the detailed imagery beamed back by New Horizons we saw a puzzling set of patterns and shapes across its surface, a series of hexagons and pentagons crisscrossing the frozen nitrogen exterior that suggested something intriguing was going on beneath. New Horizons had provided us with a hint that Pluto's heart was beating.

The geometric patterns we see on the surface of Sputnik Planitia may be distant, but they are remarkably familiar. We see these patterns all across nature, from the surface of the Sun to a pot of boiling water. They are the characteristic patterns of convection, the forms we see when a heat source causes material to rise up and

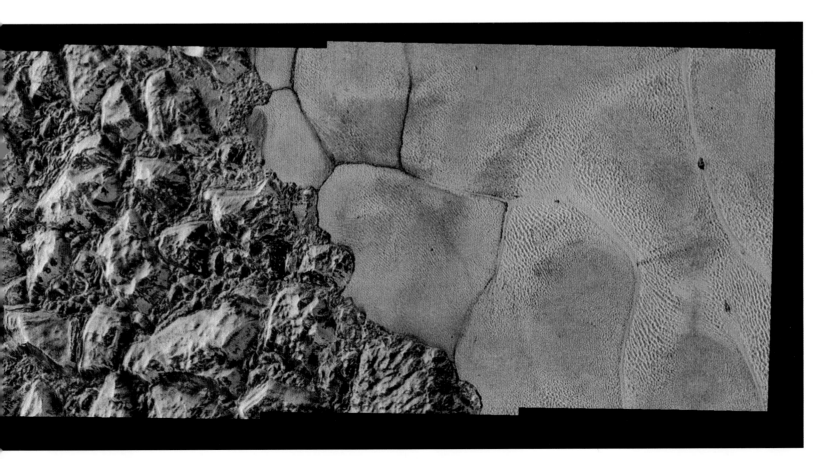

Below: This coloured image highlights the various regions on Pluto, and their diversity.

Opposite: New Horizon's flyby enabled scientists to get a better understanding of the geological processes at work in Pluto's icy world.

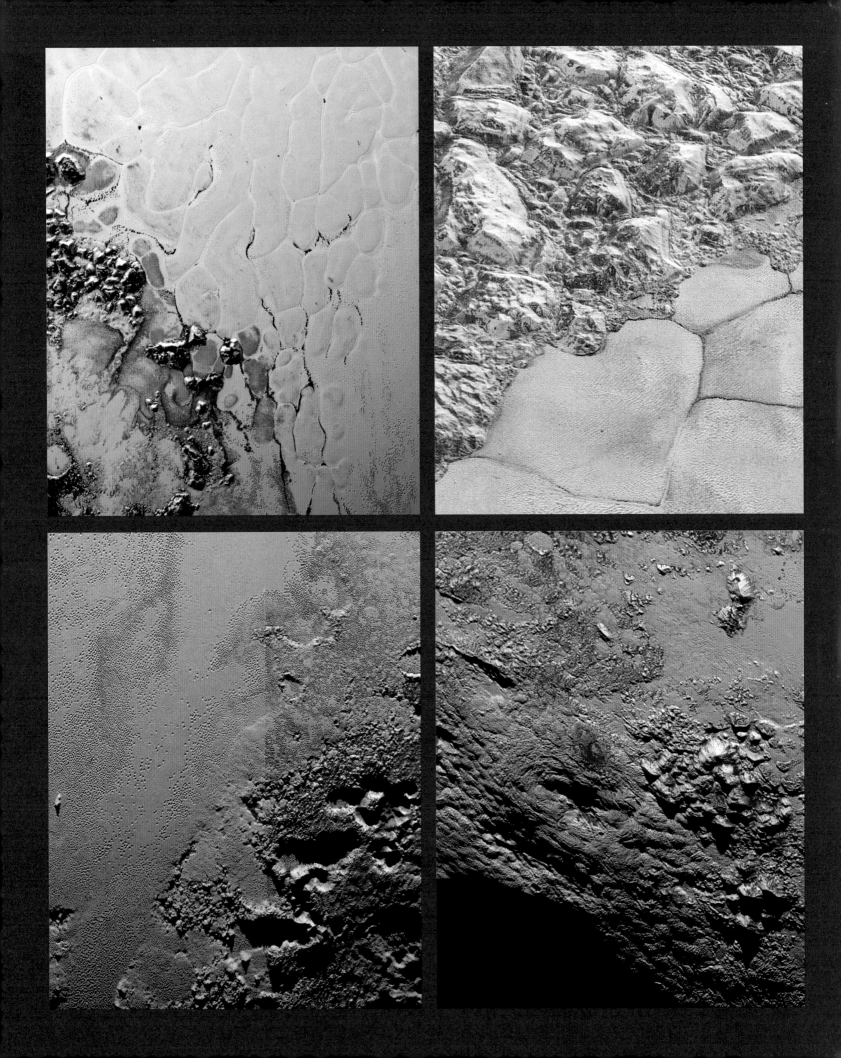

Below: The first images on the discovery of Ultima Thule, taken by the Hubble Space Telescope in 2014. On 1 January 2019, New Horizons made its flyby of this most distant target in the Solar System.

then fall back down again as it cools. So it seems that underneath the surface of this region of Pluto there is a heat source which is creating a circulating convection current, melting the nitrogen ice and causing it to rise, cool, and then fall again, constantly resurfacing the area, removing any scars from impacts and leaving the characteristic cells of convection behind.

We don't know for certain where the heat is coming from that has shaped this distant world, but many scientists now believe that below the surface of the glacier there was once a liquid body of water, a flowing ocean that might even still be there. It is this ocean that could possibly explain the strange patterns on the surface, because it needs only a small difference in temperature between the underground ocean and the frozen nitrogen above to create the convection currents needed to produce the crater-free, patterned plain we see today.

Just imagine that billions of miles away from the Sun, on the frozen frontier of the outer Solar System, there may still be an ocean of water. And what is above this lonely, distant ocean? As New Horizons left Pluto behind to continue its lonely voyage out into the darkness, NASA turned its camera back towards the dwarf planet, revealing a hazy atmosphere made mostly of nitrogen, a blue sky above an ocean of water 4.8 billion kilometres from Earth.

Today New Horizons, just like Voyager 2, continues onwards, heading out into the furthest reaches of our Solar System. In January 2019, three and a half years after leaving the beating heart of Pluto behind, it revealed a new world, the first image of 2014 MU69, otherwise known as Ultima Thule, the 'snowman' dwarf planet. By visiting these outer worlds we have momentarily illuminated the darkness, found and explored worlds far beyond our imagination and opened a new and as yet unfinished chapter in the story of our Solar System.

It's a story that began with the formation of an ordinary star a little less than 5 billion years ago, a star that would leave just the right ingredients for eight planets and countless worlds to form in her shadow. Jupiter came first, the godfather that would cast its hand across all that was to follow, shaping the destiny of the Solar System as it wandered near and far. In Mars it would leave a world too small to fulfil its potential, chasing the destiny of its nearest sibling before rapidly slowing down. Venus would soon follow. An oasis in its earliest of days, choked by the intensifying grip of its atmosphere, just like Mercury it would find itself too close to the Sun to hold on to its chance.

Only one planet, our own, has held on to its oceans beneath the thinnest of veils, a world that through the luck of its formation has not only sparked life but nurtured it as well. For 4 billion years our planet has remained stable long enough to allow the most curious of minds to evolve, look up and explore.

This is our story, a story that we now know not only has a beginning and a middle but an end, too – an end that will see the islands of life in our Solar System shift from one world to another until perhaps, in the final days of our star, it will be Saturn that harbours a world where a life form can call it a home.

We've written this story not with imagination but with exploration, a story just like the very greatest of our legends, that will be added to and heightened as it is passed down the generations to come. Where it will end up I do not know. Lost with our planet or carried onwards to distant worlds, its destiny and our destiny are unmistakably entwined.

OLD VIEW

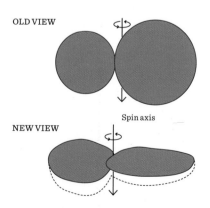

Spin axis

NEW VIEW

Left: New data has revealed that Ultima Thule is not the perfect sphere it was thought to be, rather it is a more flattened object.

Right: The most detailed images thus far received of Ultima Thule, taken in January 2019, dubbed the 'snowman' by Alan Stern, principal investigator for New Horizons.

Bottom: An early illustration of New Horizons with Ultima Thule. The spherical shape shown here is now thought to be innacurate.

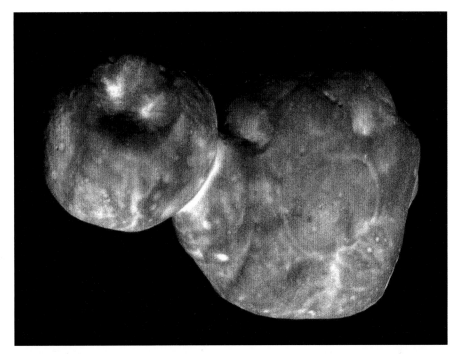

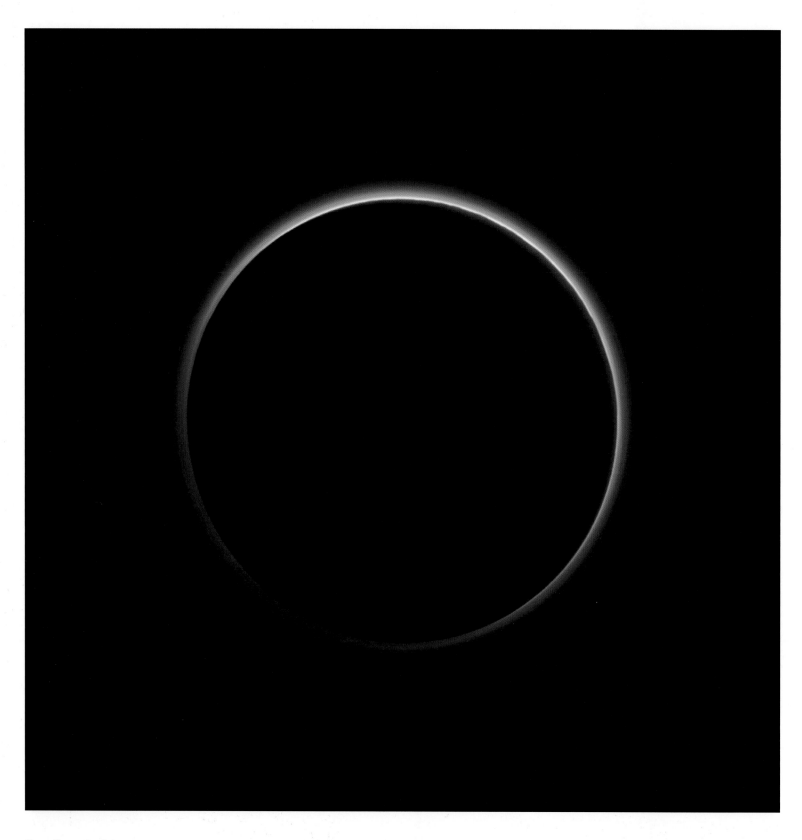

Above: The now iconic image
taken by New Horizons as it
turned its camera back one last
time when it left Pluto's orbit,
capturing the gentle glow of
the dwarf planet's atmosphere.

Why do we explore? some people ask. Shouldn't we focus on solving our problems here on Earth before spending time and energy and resources reaching for the stars?

Well, I think that focusing entirely on our mote of dust would be a profound mistake. It would mean that we've taken the decision to sit huddled in a tiny corner of the Solar System, wondering what we're doing here.

It would mean that we've taken the decision to fight among ourselves for ever more precious resources, confined below a thin shell of air to the two-dimensional surface of a small rock, rather than following the three-dimensional path marked out by the lights in the night.

We need to shift our focus. We live in a solar system of wonders, of planets of storms and moons of ice, of landscapes and vistas that stir the imagination and enrich the soul – a system of limitless resources, limitless beauty, and limitless potential. If we don't go there, we'll never go anywhere. And if we go nowhere, then we'll have no future at all.

INDEX

Page references in *italics* indicate images.

Adams, John Couch 253–4
Aegaeon 182, 210
Albion 263, 265, 269
Aldrin, Buzz 91
Anderson, Carrie 221
Andromeda galaxy 236, *236–7*, 237
Apollo missions 66, *79*, 91, 92, *92*, 93, *93*, 99
Arcturus *60*, 61
argon 105, 106, 107, 188
Ariel 248, *248*, 251, *251*
Armstrong, Neil 91, 92, *92*
Arnold, Ricky 113
Arrhenius Effect 45
Arrhenius, Svante 44, 45, *45*; *The Destinies of the Stars* 44
asteroid belt 62, 93, 123, 129, 131, 132–3, *132*, *133*, 139, 143, 147, 155, 157, 159, 160, 161, 162–3, 165, 173, 193, 235, 263, 265
asteroids 124, 129, 132, 133, *133*, 134, *134*, 135, *135*, 140–1, *158*, 159, 162–3, 165, 166, 173, 189, 216
astronomical unit 233, 235
Atlantis, space shuttle 159
Atlas 180, 182, *217*
aurora 109–10, *109*, 111, *111*, 183
autonomous guided entry system 80

Bagenal, Fran 123, 124, 247, 250, 256
Baines, Kevin 177, 191, 193, 196, 202
Banks, Sir Joseph 241
Barringer crater, Arizona 128–9, *128*, *129*, 165

Barringer, Daniel 128–9, *129*
Batygin, Konstantin 132
Beal, Jeff 129
Becker, Heidi 124, 126
Berlin Observatory 254
Big Bang 97, 109
Bouvard, Alexis 253, *253*; *Tables astronomiques* 253
brown dwarf 20
Bulletin of the Atomic Scientists: 'The Message From Mariner 4' 66
Burney, Venetia 266

Callisto 62, 63, *63*, 124, 148
Cape York meteorite 165, *165*
Cassini, Giovanni 179, 210
Cassini-Huygens mission 63, 77, 152, *152–3*, 171, *171*, 172, 177, 178, *178*, 180, *181*, 182, 183, *184*, 185, 186, *186*, 187, 191, 192–203, *192–3*, *194*, *195*, *196–7*, *198*, *199*, *200*, *201*, *202*, *203*, 205, 207, *207*, 208, 209, 210, *210*, 214, 216, 219, *219*, 220, 221, *221*, 222, *222*, 223, 224, *224*, 225, 226, *226*, 227, *228*, 229; Cosmic Dust Analyser (CDA) 207, *207*, 224, 226
Cawdron, Peter 100
Ceres 131, *131*, 132, *132*, 133, 134, *138*, 139–41, *139*, *140*, *141*, *142*, 143, *143*, 147, 155, 270, *270*
Chabot, Nancy 29, 32, 35
chaos theory/chaotic system 39, *39*, 40
Charon 265, 266, *267*, 273, *273*
Chen, Allen 77, 78
Chen-wan Yen 30

Chicxulub crater, Mexico 129, *129*, 163, *163*, *164*, 165–7, *165*
Christy, James 266, *267*
climate change 20, 50
Compact Reconnaissance Imaging Spectrometer for Mars (CRISM) 77
Cordelia 247, *247*
Crab Nebula 107, *107*
Crisp, Joy 85, 87
Curiosity rover 69, *70–1*, 77, 79, *79*, 80, *80*, *81*, 82–9, *82–8*, 95, 105, 106

Dactyl 159
d'Arrest, Heinrich 254
Data Processing Units (DPUs) 36
dating, radioactive 91, 92, 93
Dawn space probe *130*, 131, *132*, 133, 134, 139–40, 143, *143*
Deep Space Network antenna complex, Canberra, Australia 192
Desch, Professor Steve 166
Desulfurella acetivorans 58
dinosaur extinction 163, 166
Dione 180, 210, *211*, 216, *217*, 226, *226*, 227
Discovery programme, NASA 131
Discovery, Space Shuttle 109
Dougherty, Michele 187, 224, 226
dwarf planets 269–70, *270*, 271, *271*, 272, *272*, 275, 278, 279, 280
dynamo 110, 111
Dysnomia 271, *271*

Earth: atmosphere 66, 105, *105*, 108–13, *108–9*, *110*, *111*, *112*, *113*, 186, *187*, 188, *188*, 189,

190, 191, 194; aurora 109–10, *109*; colour of 257, 259; core 35, 110; dating rocks on 91, 92, 93; end of 58–61, *58–61*; formation of 121, 149, 153, 155–7, 175, 189; great oxygenation event 189; greenhouse effect and 19, 20, 23, 45, 55–7, *55*; habitable nature of 2, 20, 25, 58–9, 157, *157*, 186, 223, 279; hydrothermal vents/activity 99; Jupiter and 123, *123*, 124, *129*, 133, *133*, 134, 149, 153, 155, 163, *163*, *164*, 165–7, *165*; Karman line 186; life begins on 97, 98–9, 226, 227, *227*; life cycle of Sun and 20, *21*, 23, 55; Mars and 66, 69, 72, *74*, 75, *75*, 77, 80, 87, 89, 91, 95, 97, 98–9, 100, 103, 105, 106, 108–13, *108–9*, *110*, *111*, *112*, *113*, 115, 117; Mercury and 27, 29, 30, 35, 39, 40, 41; meteorites and 92, 129, *129*, 133, *133*, 134, 163, *163*, *164*, 165–7, *165*; Moon and 35, 39, 66, 72, 81, *90*, 91, 92–3, *92*, *93*, 95, 98, 111, 133, 149, 163, 213, 214, *215*, 233; Neptune and 255, 256, 259, 261; protective shield/magnetic field 108–13, *108–9*, *110*, *111*, *112*, *113*; Saturn and 171, 175, 176, *176*, 177, 182, 183, 185, 190, 191, 203, 204; Uranus and 242, 244, 246, 248; Venus and 44, 45, 46, 49, 50–1, 53, 55, 56, 57, 61; 'von Karman vortex street' 186, *187*
Earthrise 66, 91
Ecclesiastes 1:9 20
Ehlmann, Bethany 139, 141, 143
Eigenbrode, Jennifer 72
Einstein, Albert 29, 169
Eliot, T. S. 99
Enceladus 62, *62*, 63, 171, 180, 182, 194, *217*, 224–9, *224*, *225*, *226*, *227*, *228*
Encke, Comet 40
Encke, Johann Franz 40
Encounters at the End of the World (film) 115
Energetic Particle and Plasma Spectrometer (EPPS) 36

Epimetheus 180, *181*, *217*
Eris 269–70, *270*, 271, *271*
ESA (European Space Agency) 77, 119, 136, 147, 148, 153, 166, 219
Europa 62, 63, 124, 148, 149, 159
evolution by natural selection 60, 96, 99, 115, 157
exoplanets 61, 144–5, 175

Flando, Gary 239
Fletcher, Leigh 119, 136, 147, 148, 153, 156, 166, 199

Gale, Walter Frederick 84
Galilei, Galileo 62, 178, 253
Galileo Jupiter spacecraft *122*, 123, 124, 125, *125*, *146*, 147, 149, 151, *151*, 159, *159*, 160, *160*, 161, *161*
Galle, Johann 254
Gamma-Ray and Neutron Spectrometer (GRNS) 36
Ganymede 62, 63, *63*, 124, 148, 149, *149*
gas giants 62, 144, 145, 147, 152, 155, 156, 160, 169, 171, 172, 173, 175, 176, 191, 194, 200, 204, 232, 241, 244, 245, 269 *see also individual planet name*
Gaspra 159
Gauss, Friedrich 139
Genio, Anthony Del 53
George III, King 241
Gliese system 144–5, *144*, *145*, 147
Goddard Institute for Space Studies (GISS), NASA 50, 53
Godfrey, David 185
Goldilocks zone 20, 41, 54, 186, 223
Grand Tack Hypothesis 41, 131, 143, 144, 147, 152–7
Grand Tour 239
greenhouse effect 19, 20, 23, 45, 55–7, *55*

Grinspoon, David 20, 23, 49, 55, 56, 61, 223, 239, 242, 244, 255, 259, 261, 267
Guidice, Rick 57

habitable zone 45
Haleakala National Park, Hawaii *38*, 39
Hall, Asaph 194, 196
Halley, Edmond 233, 235
Hammond, Noah 262
Haumea 269, *270*
HED meteorites 133, *133*, 134
Herschel, John 247
Herschel, William 179, 210, 240–1, 247, 267
Hooke, Robert 179
Horrocks, Jeremiah 51
hot jupiters 145
Hubble Space Telescope 23, 66, *66*, 77, 78, 139, 160, 233, 255, 257, 266, 267, *267*, 274, 278
Huygens, Christiaan 63, 178–9, 183, 219, 223
hydrothermal vents 58, 98, 99, 226, *226*, *227*
Hyperion 180, 210, *213*, 213, *217*

Iapetus 180, 210, 213, *213*, *217*
ice giants 173, 237, 244–5, 253, 256, 257, 259 *see also individual planet name*
Ida 159
Innaarsuit, Greenland 112
InSight lander *70–1*
International Astronomical Union (IAU) 235, 269–70
International Space Station (ISS) 105, 109, 113
Inuit people 165
Io 62, 63, *63*, 124, *125*, 147, 148, 149, 150, *150*, 151, *151*, 159, 239, 265
Iridium 165
isochron method 91
Isua, Greenland 91, 99

James Webb Space Telescope 78

Janus 180, *181*, 182, *182*, *217*

Jet Propulsion Laboratory (JPL), NASA 30, 68, 77, 87, 115, 124, 183, 200, *200*, 239, 260

Johnson, Lyndon B. 66, 69, *69*, 77, 78

Juno (asteroid) 134, 139

Juno mission 124, 125, 126, *126*, *127*, *154*, 166, *167*; JunoCam 125, 126, *126*, *127*

Jupiter 39, 41, 62–3, 118–67, 172, 173, 278; asteroid belt and 129, 131, 132–3, *132*, *133*, 134, *134*, 135, *135*, *138*, 138, *139*, 140–1, *140*, *141*, *142*, 143, *143*, 155, 157, 159, 161, 162–3, 165; atmosphere 123, 124, 136, 154, *154*, 159, 161, 166, *166*, 183, 188, *188*; Cassini probe and 152, *152–3*, 193, 194, *194*; core 121, 177; formation of 120, 121, *121*, 124, 145, 147; Galileo Jupiter spacecraft and *122*, 123, 124, 125, *125*, *146*, 147, 149, 151, *151*, 159, *159*, 160, *160*, 161, *161*; gas giant 145, 147, 176, 235, 244; Grand Tack Hypothesis/change in orbit 41, 131, 143, 144, 147, 152–7; gravitational field/effect upon other planets of 124, 128–9, *128*, *129*, 147, 148, 149, *149*, 150, *150*, 151, *151*, 152–7, *152–3*, *154*, *156–7*, 159–66, *160*, *161*, *163*, *164*, 165–7, *165*, 279; Great Red Spot 124, *124*, 126, *126*, *136*, 137, *137*, 157, *157*, 239, 253, 255; Icy Moons Explorer mission and 119, 136, 147, 148, 153, 156, 166; Juno mission and 124, 125, 126, *126*, *127*, *154*, 166, *167*; mass 121, 123, 188, *188*; meteorite impacts on Earth and 121, 128–9, *128*, *129*, 134, 159, 163, *163*, *164*, 165–7, *165*; moons 62, *62*, 63, *63*, 119, 124, 147, 216, 239, 265; Pioneer missions and 123, 124, *124*, 125, *125*, 165, *165*; Shoemaker-Levy 9 and 159, 160–1, *160*, *161*; size of 121, 123, *123*, 148, 149, 153, 155, 177, 188, *188*, 191; South pole 124, *124*, 166,

167, *252*, 253; storms/weather systems 126, *126–7*, *136*, 137, *137*, 152, *152–3*, *154*, *156–7*, 157, 159; Voyager probes and 123, 149, *149*, 150, *150*, 151, *151*, 161, *238*, 239, 241, 242, 260; water on 124

Kamchatka Peninsula, The, Eastern Siberia 58, *58*, *59*, 60

Kármán, Theodore von 186

Keck telescope, Hawaii 144, *144*

Kelly, Scott 105

Kepler, Joseph 233

Kepler Space Telescope 144, *144*

K-Pg boundary 165

Kuiper Airborne Observatory 246

Kuiper Belt 93, 216, 237, 263, *264–5*, 265, 267, 269, 270

Kuiper, Gerard 260

Lalande, Jérôme 235

Laplace, Pierre-Simon 180

Lassell, William 251, *251*, 260

Late Heavy Bombardment 93, 99, 100, 111, 155, 163, 189

Le Verrier, Urbain 254, 267

Levy, David 159

light absorption spectrum 256, 258, *258*

Lowell Observatory, Flagstaff, Arizona 266

Lowell, Percival 69, 266

Luna missions 92, 93, *93*

Lunine, Jonathan 171, 210

Madagascar *113*

Magellan mission 50, *52*, 53, *53*

Magnetometer (MAG) 36

main sequence star 55, 121, *121*, 144

Maize, Earl 195

Makemake 269, 270, *270*

Mangalyaan space probe 77

Mariner 4 66, *67*, 68, *68*, 69, 115

Mariner 9 69, 72, 100

Mariner 10 *26*, 27, *27*, 28, *28*, 31, *31*, 32, 40, 42, *42*, 43, *43*

Mars 23, 25, 39, 41, 49, 51, 58, 61, *65*, 66–117, 155; Aeolis region 97, *97*; Amazonian period 75, *75*, 93, 103; Amazonis Planitia region 72, 75; Arabia Terra region 95, 96, 97; atmosphere 69, 75, 77, 79, 80, 96, 97, 100, 102–3, 105, 188, *188*; atmosphere, loss of original 104–11, *104*, *105*, *106*, *107*, *110*, 186; aurora 111, *111*; climate 75, 96–7, *96–7*, 98–9, *98*, 100, 105, 106, *106*, 111; colonies on 51, 75, 115, 117; crater counting on 93, 95; Curiosity rover and 69, *70–1*, 77, 79, *79*, 80, *80*, *81*, 82–9, *82–8*, 95, 105, 106; death of 100–3, *100*, *101*, *102*, *103*, 111; dynamo 111; Echus Chasma 100, 102–3, *102*; Elysium Planitia 70, *70–1*; Eridania Sea 98–9, *98*, *99*; ExoMars rover 77; ExoMars Trace Gas Orbiter 77; formation of 72, 75, *75*, 99, 111, 153, 175, 279; Gale crater 69, 80, *80*, 84–5, *84–5*, 87, 88, *88*, 95; Hellas Basin 72, 100, 101, *101*, 111; Hesperia Planum 72, 100, 111; Hesperian Period 75, *75*, 84, 88, 93, 95, 96, 100, 102–3, 111; InSight lander and *70–1*; Jezero crater 95, *95*; Jupiter grand tack and 147, 149, 153, 155, 279; Kasei Valles 103; life on 66, 68, 69, 75, *75*, 77, 78, 88, 97, 98–9, 102, 103, 108, 109, 110, 115; magnetic field 111; Mangalyaan space probe and 77; manned missions to 115, 117; maps, reading 70, *70–1*, 72–5, *73*, *74*, *75*; Mariner 4 and 66, *66*, *67*, 68, *68*, 69, *69*, 115; Mariner 9 and 69, 72; Mars Atmosphere and Volatile EvolutioN Mission (MAVEN) spacecraft and 75, 77, *104*, 105, 106, *106*, 107, 110, 111;

Mars Express and 77, 111; Mars Global Surveyor and 70, 72, 96–7, 98; Mars Odyssey and 77; Mars Pathfinder and *70–1*, 79; Mars Reconnaissance Orbiter (MRO) and *68*, 69, *76*, 77, *94*, 95, 98, 100, 105; Mars Science Laboratory mission and 77–8; Meridiani Planum 73, *73*, 77; mineral resources 115, 117; Noachian Period 72, 75, *75*, 84, 88, 93, 95, 96, 98, 100; Noachis Terra region 72; Noctis Labyrinthus 103, *103*; Olympus Mons 72, 100, *100*, 134; Opportunity rover and *70–1*, 77, *77*, 79, *79*; Phoenix lander and *70–1*; relative timelines of life on Earth and activity on 75, *75*; size of 97, 103, 111, 188, *188*; South Pole 106, *106*; Spirit rover and *70–1*, 79, *79*; Tharsis Rise 72, 100, 101, *101*, 102, 103; topography of 66, *66*, 70, *70–1*, 101, *101*; Tyrrhenus Mons 100, 103, *103*; Utopia and Chryse Basins 72; Valles Marineris 72, 103, *103*; Viking landers and 69, 70, *70–1*; water on 49, 66, 69, 73, *73*, 75, 77, 82, *82*, 84, 87, 88, 95–7, *95*, *96–7*, 98–9, *98*, *99*, 100, 102, *102*, 103, *103*, 105, 106, *106*, 111, 115
Mars Color Imager (MARCI) 77
Mars Orbiter Laser Altimeter (MOLA) *70–1*, 98, 101, *101*
Maskelyne, Nevil 240–1
Mauna Kea Observatories, Hawaii 144, *144*
Maxwell, James Clerk 180; *On the Stability of the Motion of Saturn's Rings* 180
Merbold, Ulf 105
Mercury 23, 24, 25, 26–43, 44, 46, 58, 61, 172; anatomy of 40, *40*; atmosphere 27, 29, 31, *34*, 35, 39, 42, *42*, 188, *188*, 278; Caloris Planitia 32–3, *32*, *33*, 37, *37*; core 27, 35, 40, *40*, 41; craters on 27, 28, *28*, 30, *30*, 31, 32–3, *32*, 35, 36, 37, *37*, 41, *41*; formation

of 35, 40, 41, 153; geology of 27, 30, *30*, 33, *33*, 37, *37*, 40, *40*; magnetic field 27, 35, 40–1; mapping 32–7, *32*, *33*, *34*, *35*, *36*, *37*; Mariner 10 and *26*, 27, *27*, 28, *28*, 31, *31*, 32, 40, 42, *42*, 43, *43*; orbit of 27, 29, *29*, 30, *30*, 31, 35, 39, 40, 41, 46, 58, 145; size of 40, *40*, 171, 186, *186*, 188, *188*, 219; Sun death and 61; Voyager 1 and 41, *41*; water ice on 30, *30*, 33, *33*, 35; wrinkle ridges on 40, *40*
Mercury Atmosphere and Surface Composition Spectrometer (MASCS) instrument, The 35
Mercury Dual Imaging System (MDIS) *34*, 35, 36
Mercury Laser Altimeter (MLA) 36
Messenger mission 29, *29*, 30–1, *30*, *31*, 32, 33, *34*, 35, *35*, 36, *36*, 37, *37*, 40–2, *41*
Meteor crater (also known as the Barringer crater), northern Arizona desert 128–9, *128*, *129*, 165
Meteor Crater Enterprise 129
Meteorite ALH 84001 111
metre-size barrier 25
Meyer, Michael 88
Milky Way *116–17*, 117, 152, 157, 232, 236
Mimas 171, 180, 206, *206*, 210, *216*, *217*
Miranda 248, *248*, 250, *250*
Moon 35, 39, 66, 72, 81, *90*, 91, 92–3, *92*, *93*, 95, 98, 111, 133, 149, 163, 213, 214, *215*, 233
Murray, Carl 216, 229
Musk, Elon 117

NASA *see individual programme and section*
National Optical Astronomy Observatories, Tucson, Arizona 185
Near-Earth Objects 159
Neptune 93, 161, 173, 216, 232, 233, *233*, 237, 239, 244, 245, 252–65, 266; anatomy of 256, *256*; atmosphere 188, *188*, 253,

255, *255*, 256, 257, 259; colour 257, 259; core 256, *256*, 257; discovery of 233, *233*, 253–4, *255*; formation 245, 256, 257; geysers on 262, *262*, 263, 265; Great Dark Spot 253, *253*, 254, 255, 257, *257*; ice giant 257; landscape 258, *259*; moons 253, 255, 260–3, *260*, *261*, *262*, *263*, *264*, 265; rotation 257, *257*, 261; size of 253; South pole *252*, 253; storms 253, *253*, 255–6, 257; Voyager probes and 235, 239, 253, 254–5, 256, 259, 260, 261–2, 263, 265
Nereid 253, 255, 260
neutrinos 109
New Horizons mission 123, 124, 247, 250, 256, 263, 268–80, *268*, *269*, *270*, *272*, *273*, *274–5*, *276*, *277*, *278*, *279*, *280*
Newton, Isaac 29; laws of universal gravitation 39, 253
NGC 2362 (star cluster) 175, *175*
Nittler, Larry 29, 30–1, 39, 49
Nix 273, *273*
nuclear fusion 55, 109
Nuvvuagittuq crustal belt, Quebec 99

Oberon 247, 248, *248*, 251, *251*
Olbers, Heinrich 134
Open University 95
Ophelia 247
Opportunity rover *70–1*, 77, *77*, 79, *79*
orbital resonance 29, 149
Orion Nebula *120*, 121
Oschin Schmidt Telescope, San Diego, California 145, *145*

Pallas (asteroid) 134, 139
Palomar Observatory, California 159, 269
Pan 210, *210*, *217*
Pandora 182, *217*
Pappalardo, Bob 169, 172, 208, 214

Pathfinder, Mars *70–1*, 79

Phoenix lander *70–1*

phytoplankton 112, *112*

Piazzi, Giuseppe 139

PICA (phenolic impregnated carbon ablator) 80

Pickering, William H 69, *69*

Pioneer 10 123, 125, *125*, 165, 237

Pioneer 11 123, 124, *124*, 180, *181*, 182

Pioneer Venus Orbiter 44, *44*, 49, 50, 57

planetary classification 269–70

Planetary Science Institute 20, 55

Pluto 232, 237, 265, 266–70; atmosphere 188, *188*, 278, 280, *280*; beauty of 274–5; classification as a planet 269–70; discovery of 265, 266–7, *266*, 269; geology/patterns of convection 274, *274*, 275, 277, *277*, 278; moons 266, *266*, 269, 270, *270*, 271, *271*, 273, *273*; New Horizons mission and 123, 124, 247, 250, 256, 263, 268–80, *268*, *269*, *270*, *272*, *273*, *274–5*, *276*, *277*, 278, *279*, *280*; orbit 237, 266, 269, 270, 280; regions of 275, *275*, *276*, *276*, 278; surface 266, *266*, 274–5, *274–5*, *276*, *277*, 278

Porco, Carolyn 183, 185, 199

Proctor, Richard 44

Prometheus 182, 217, *217*

Proton gradients 98–9

protoplanets 20, 25, 134, 173, 175

Pullman, Philip: *The Subtle Knife* 111

pulsar 24–5, *24–5*

quantum mechanics 109

Quaoar 269, 271, *271*

red giant 20, 23, 61

relativity, theory of 29

Rhea 180, *216*, *217*

Rivera, Eugenio 144

Robert C. Byrd Green Bank Telescope, West Virginia 211

Roche, Edouard 213, 214

Roche limit 160, 213, 214

Ross 248, 237

Rothschild, Lynn 19

Royal Greenwich Observatory 254

Royal Society 179, 240, 241

rubidium-strontium dating 91

Sagan, Carl 68, 69

Salar de Uyuni, Bolivia *138*, 139

Saturn 62, 63, 155, 168–229, 232, 237, 253, 278; anatomy of 191, *191*; atmosphere 171, 175, *175*, 177, 183, 185, 186, *186*, 187, *187*, 188, *188*, 189, 190–1, *190*, *191*, 192, *192*, 193, 194, 199, 200, 202–3, 207; Cassini probe and 80, 171, *171*, 172, 177, 178–9, *178*, 180, *181*, 182, 183, 184, *184*, 186, *186*, 187, 191, 192–203, *192–3*, *194*, *195*, *196*, *197*, *198*, *199*, *200*, *201*, 202, *203*, 205, 207, *207*, 208, 210, *210*, 214, 216, 219, *219*, 220, *220*, 221, *221*, 222, *222*, 223, 224, *224*, *225*, 226, *226*, *228*, 229; climate 180, 202; core 171, 176, 177, 185, *185*, 191, *191*, 203; formation of 172–3, 174, *174*, 175, 176–7, *176–7*; gas giant 62, 155, 171, 172, 173, 175, *175*, 176, 180, *181*, 182, *182*, 191, 194, 200, 204, 232, 244; ice line and 173, *173*, 175; mass 171, 175, 188, *188*, 190, 191, 203, 241; moons 63, 169, 171, 178, 179, 180, *181*, 182, *182*, 183, 193, 194, 202, 206, *206*, 208, 210–29, *210–29*; North polar region *184*, 185, *185*, 198, *198*; orbit 155, 177, 180, 190, 210; Pioneer 11 and 123, 180, *181*, 182; rings *170*, 171, 172, *172*, 175, *175*, 178–9, *178*, *179*, 180, 182, *182–3*, 183, 192, 193, 194, 200, 202, *203*, 204–9, *204–5*, *206*, *207*, *208*, *209*, 210, 214, 216, 217, *217*, 229, 242, 246; size of 171, 175, 177, 190, 191; storms 171, 175, *175*, 185, 194, 196, *196–7*, 199, *199*, 202–3, 255; Voyager probes and 182, 183, 185, 193, 194, 198, 199, 203, 210, 235, *238*, 239

Saturn Orbital Insertion (SOI) manoeuvre 193, 194, *194*

Sedna 269, 270, *270*

Shakespeare, William: *Macbeth* 19

shepherding moons 182, 210

Shoemaker, Carolyn 159

Shoemaker, Eugene 129, 159

Shoemaker-Levy 9 159, 160–1, *160*, *161*

Sirius A *60*, 61

Sirius B *60*, 61

solar day 29, 51

Solar Dynamics Observatory 22

solar radiation 20, 27, 55

Solar System: age of 92; formation of 25, 40, 41, 72, 95, 97, 120, 121, 129, 131, 132, 134, 144, 145, *145*, 156; Jupiter shaping of *see* Jupiter; odd nature of 144; scale of 51, 232, *232*, 233, 235, *236–7*, 237; as a system 98, 108; unstable nature of 41, 54, 223

Solomon, Sean 29

Spilker, Linda 205, 208, 229

Spirit rover *70–1*, 79, *79*

Standard Iron Company 129

Stedman, M.L. 32

Steltzner, Adam 79

Stern, Alan 231, 263, 269, 273, 274, 279

stromatolites 189, *189*

S/2004 S6 and S/2004 S3 210

Sun: corona 108; formation 25, 120, 121; greenhouse effect and 55, 56–7, 61;

Jupiter and 120, 121, 123, 124, 132, 143, 147, 152, 153, 162, 163, 165; life cycle of 20, *20*, *21*, 23, 25, 54, 55, 61, 121, *223*; Mars and 69, 97, 106, 108–11; Mercury and 27, 29, *29*, 30–1, 35, 36, 39, 41, 58; Neptune and 255, 256, 259, 262; Pluto 269, 270, 275, 278; red giant phase 20, *20*, *21*, 23, 61; Saturn and 171, 173, 175, 177, 183, 185, 186, 190, 191, 194, 199, 204; solar activity 22, *22*; solar prominence 109, *109*; solar wind 77, 105, 106, 108, 109, 110, 111, 186, 237; Uranus and 237, 240, 242, 244, 248; Venus and 40, 50, *50*, 51, 53; weather, as driver of 194
Super Earths 144–5, *144*, *145*, 152, 155
supernovae 121

Theia 91, 189
thermal radiation 55
tidal heating 148, 149, 265
Titan 63, *63*, 169, 171, 178, 180, *181*, 183, *183*, 194, 210, *212–13*, 213, *216*, *217*, *219–23*, *218–23*, 229, *229*, 239
Titania 247, 248, *248*, 251, *251*
Tolkien, J.R.R.: *The Fellowship of the Ring* 119
Tombaugh, Clyde 266–7, *266*, 269
Triton 253, 260–3, *261*, *262*, 265
T Tauri-type star 121
Tungurahua Volcano, Ecuador 99
Type II migration 145

Ultima Thule 278, *278*, 279
Ulysses spacecraft 160
Umbriel 248, *248*, 251, *251*

University College London 95
Uranus 93, 216, 232, 237, 239, 240–51, 253, 257, 266; anatomy of 242, *242*; atmosphere 188, *188*, 242, 243, *243*, 244, *244*, *245*, 246, 248, *249*, 259; axial tilt 248; climate 242; colour of 259; core 242, *242*, 244, 257; discovery of 240–1; formation of 244–5, 246, 248, 256, 263; ice giant 173, 238, 244, 245, *245*, 256; mass 188, *188*, 242–3; moons 241, 242, 247, 248, *248*, 250, *250*, 251, *251*; orbit 238, 253, 254, 266; rings 241–2, 246–7, *246*, *247*; rotation 51, 242, 243, *243*; size 145, 244; Voyager probes and 235, 239, 240, *240*, 241–2, *243*, 244, *244*, 246, 247, *247*, 248, 251

Vasavada, Ashwin 80, 102, 115
Venera programme 45, *45*, 46, *46*, 47, *47*, 49, 50
Venus 23, 41, 44–57, 58, 61; atmosphere 42, *43*, 44, 45, 46, 49, 53, 55, 56–7, 186, 188, *188*, 189, 194; Cassini and 193, 194, *194*; climate 19, 23, 45, 46, 50, 51, 53, 55–7, 58; Earth and 44, 45, 46, 49, 50–1, 53, 55, 56, 61; formation of 25, 48–53, 153, 175; Galileo spacecraft and 159; greenhouse effect on 19, 23, 55–7; life on 44, 45, 50–1, 53, 55; Maat Mons *48–9*, 49; Magellan mission and 50, *52*, 53; Mariner 10 and 27, 42, *43*; Messenger spacecraft and 27, 30, *30*; mysterious nature of 44; place in the Solar System 55; rotation on axis/ sidereal day 51, 52, *52*; size of 44, 46, 49, 53, 188, *188*; Sun life cycle and 55,

61; surface of 44–5, *44*, 46, *46*, 49, 50, 53, 55, *56*, 57, *57*; transit of 50, *50*, 51, *51*, 235, *235*; Venera programme and 45–6, *45*, *46*, 47, *47*, 49, 50; water on 44, 45, 46, 49, 50–1, 53, 55, 56–7
Veritas 213, 226
Vesta 131, *131*, *132*, 133, *133*, 134, *134*, 135, *135*, 139, 147, 155
Viking landers 69, 77, 79, 80, 88
Voyager probes 77; Voyager 1 41, *41*, 123, 149, *149*, *150*, 151, *151*, 182–3, 193, 194, 203, 224, 237, 239; Voyager 2 123, 149, *149*, 151, *151*, 161, 182, 183, 185, 193, 194, 198, 199, 210, *234*, 235, 237, *237*, *238*, 239, 240, *240*, 241, 242, 243, *243*, 244, *244*, 246, *246*, 247, *247*, 248, 251, *251*, 253, 254–6, *254*, 259, 260–3, *261*, *262*, 265, 278
VVEJGA (Venus-Venus-Earth-Jupiter Gravity Assist) trajectory 193

Wallace, Alfred Russel 69
weak nuclear force 109
Wells, H.G.: *The War of the Worlds* 66
Weywot 271
white dwarf 20, 23, *23*, 61, 63
Wordsworth, Robin 96, 97
wrinkle ridges 40

X-Ray Spectrometer (XRS) 36

Ypey, Nicholas 50, *50*

Zubrin, Robert 117

PICTURE CREDITS
t: top, b: bottom, l: left, r: right, m: middle

All reasonable efforts have been made to trace the copyright owners of the images in this book. In the event that there are any mistakes or omissions, updates will be incorporated for future editions.

1 Shutterstock; 2 JEFF DAI / SCIENCE PHOTO LIBRARY; 5 Shutterstock; 7 Shutterstock; 8 BABAK TAFRESHI / SCIENCE PHOTO LIBRARY; 10 t Science History Images / Alamy Stock Photo, b Pluto / Alamy Stock Photo; 11 SPUTNIK / SCIENCE PHOTO LIBRARY; 12 PLANETARY VISIONS LTD / SCIENCE PHOTO LIBRARY; 13 NASA / SCIENCE PHOTO LIBRARY; 14 NASA/JPL-Caltech/Univ. of Arizona; 15 NASA Earth Observatory images by Joshua Stevens; 16 NASA / SCIENCE PHOTO LIBRARY; 19 Shutterstock; 20 Fsgregs Wikimedia commons; 21 DETLEV VAN RAVENSWAAY / SCIENCE PHOTO LIBRARY; 22 NASA/GSFC/SDO; 23 t MSFC, b NASA, ESA, and K. Noll (STScI); 24 NASA/JPL-Caltech; 25 NASA/JPL-Caltech; 26 t Shutterstock, b NASA/Johns Hopkins University Applied Physics Laboratory/Carnegie Institution of Washington; 27 NASA / SCIENCE PHOTO LIBRARY; 28 tl NASA/JPL, tr NASA, COLOURED BY MEHAU KULYK / SCIENCE PHOTO LIBRARY; b NASA Wikimedia Commons; 29 t HarperCollins, b NASA/Johns Hopkins University Applied Physics Laboratory/Carnegie Institution of Washington; 30 t and bl HarperCollins, br NASA/Johns Hopkins University Applied Physics Laboratory/Carnegie Institution of Washington; 31 t NASA/Johns Hopkins University Applied Physics Laboratory/Carnegie Institution of Washington, b NASA/JPL; 32 NASA/Johns Hopkins University Applied Physics Laboratory/Carnegie Institution of Washington; 33 NASA/Johns Hopkins University Applied Physics Laboratory/Carnegie Institution of Washington; 34 NASA/Johns Hopkins University Applied Physics Laboratory/Carnegie Institution of Washington; 35 NASA/Goddard Space Flight Center Science Visualization Studio/Johns Hopkins University Applied Physics Laboratory/Carnegie Institution of Washington; 36 NASA/Johns Hopkins University Applied Physics Laboratory/Carnegie Institution of Washington; 37 NASA/Johns Hopkins University Applied Physics Laboratory/Carnegie Institution of Washington; 38 WALTER PACHOLKA, ASTROPICS / SCIENCE PHOTO LIBRARY; 39 SCOTT CAMAZINE / SCIENCE PHOTO LIBRARY; 40 t NASA/Johns Hopkins University Applied Physics Laboratory/Carnegie Institution of Washington, bl MARK GARLICK / SCIENCE PHOTO LIBRARY, br NASA/Johns Hopkins University Applied Physics Laboratory/Carnegie Institution of Washington; 41 t NASA/Johns Hopkins University Applied Physics Laboratory/Carnegie Institution of Washington, m redrawn from NASA/Johns Hopkins University Applied Physics Laboratory/Carnegie Institution of Washington, b NASA/Johns Hopkins University Applied Physics Laboratory/Carnegie Institution of Washington; 42 US GEOLOGICAL SURVEY / SCIENCE PHOTO LIBRARY; 43 NASA / SCIENCE PHOTO LIBRARY; 44 l NASA/ARC, r MARK GARLICK / SCIENCE PHOTO LIBRARY; 45 Science History Images / Alamy Stock Photo; 46 t Shutterstock, b SPUTNIK / SCIENCE PHOTO LIBRARY; 47 tl SPUTNIK / SCIENCE PHOTO LIBRARY, tr Sovfoto/UIG via Getty Images, b (all) SPUTNIK / SCIENCE PHOTO LIBRARY; 48 NASA/JPL; 50 t United States Naval Observatory, b LIBRARY OF CONGRESS / SCIENCE PHOTO LIBRARY; 51 t United States Naval Observatory, b NASA / GODDARD SPACE FLIGHT CENTER / SDO / SCIENCE PHOTO LIBRARY; 52 NASA/JPL/USGS; 53 NASA; 54 frans lemmens / Alamy Stock Photo; 55 t NASA/JPL-Caltech, Illustrations by Jessie Kawata; b HarperCollins; 56 NASA; 57 NASA; 58 Igor Shpilenok / naturepl.com; 59 Igor Shpilenok / naturepl.com; 60 t DSS2 / MAST / STScI / NASA, b NASA, ESA and G. Bacon (STScI); 61 Science History Images / Alamy Stock Photo; 62 NASA/JPL-Caltech/SETI Institute; 63 t NASA/JPL/University of Arizona/University of Idaho, b NASA/JPL/DLR; 65 Shutterstock; 66 NASA and the Hubble Heritage Team (STScI/AURA); 67 NASA Image Collection / Alamy Stock Photo; 68 tl NASA/JPL, tr NASA/JPL-Caltech/Univ. of Arizona, b NASA/JPL-Caltech/Dan Goods; 69 NG Images / Alamy Stock Photo; 70 NASA/JPL-Caltech; 73 NASA/JPL-Caltech/Univ. of Arizona, tr; 74 B&M Noskowski / Getty Images; 75 HarperCollins, drawn from information supplied in 'The Climate of Early Mars', Robin D Wordsworth, Annual Review of Earth and Planetary Sciences 2016. 44: 1–31; 76 JPL/NASA; 77 NASA/JPL-Caltech; 78 t NASA/JPL-Caltech/Univ. of Arizona, b Bill Ingalls / NASA / Handout / Getty Images; 79 t Redrawn from NASA/JPL-Caltech, b NASA/JPL-Caltech/Cornell Univ./Arizona State; 80 NASA/JPL-Caltech/ESA/DLR/FU Berlin/MSSS; 81 tl NASA/JPL, Pioneer Aerospace, tr NASA/JPL-Caltech, bl and br NASA/JPL-Caltech; 82 NASA/JPL-Caltech; 84 NASA/JPL-Caltech/MSSS; 85 t NASA/JPL-Caltech/MSSS; 86 NASA/JPL-Caltech/MSSS; 87 NASA/JPL-Caltech; 88 t HarperCollins, m and b NASA/JPL-Caltech/MSSS; 89 Danita Delimont / Getty Images; 90 NASA; 92 t NASA / Johnson Space Center, bl NASA / Marshall Space Flight Center, br NASA / JSC; 93 Redrawn from Lunar Sourcebook: A User's Guide to the Moon by Grant H. Heiken, David T. Vaniman, Bevan M. French, image courtesy NASA; 94 NASA/JPL-Caltech/Univ. of Arizona; 95 NASA/JPL-Caltech/Univ. of Arizona; 96 USDA / FSA; 97 t NASA/JPL/Malin Space Science Systems; 98 HarperCollins; 99 t NASA, b Ammit / Alamy Stock Photo; 100 NASA/JPL-Caltech/University of Arizona; 101 NASA/JPL; 102 NASA/JPL-Caltech/Univ. of Arizona; 103 t NASA/JPL-Caltech/Arizona State University, b NASA/JPL-Caltech/Univ. of Arizona; 104 NASA/Bill Ingalls; 105 NASA / Scott Kelly; 106 l NASA/Univ. of Colorado, r NASA/JPL-Caltech/Univ. of Arizona; 107 NASA/ESA/JPL/Arizona State University; 108 NASA / Johnson Space Center; 109 t NASA/SDO; 110 t HarperCollins, b NASA/GSFC; 111 Reprinted from Elsevier, Vol 115, J. Lilensten, D. Bernard, M. Barthélémy, G. Gronoff, C. SimonWedlund, A. Opitz, 'Prediction of blue, red and green aurorae at Mars', Pages 48–56., Copyright 2015, with permission from Elsevier; 112 t NASA/U. S. Geological Survey/Norman Kuring/Kathryn Hansen, b NASA/USGS/Joshua Stevens/Kathryn Hansen; 113 NASA; 114 NASA/JPL-Caltech, Illustrations by Invisible Creature; 116 Mike Mackinven / Getty Images; 119 Shutterstock; 120 Heritage Space/Heritage Images/Getty Images; 121 t NASA/STScI, b NASA/GSFC; 122 Roger Ressmeyer/Corbis/VCG via Getty Images; 123 DAMIAN PEACH / SCIENCE PHOTO LIBRARY; 124 SSPL/Getty Images; 125 tl ARC-1973-AC73-9341, tr NASA/JPL-Caltech/SETI Institute, mr Space Frontiers/Hulton Archive/Getty Images, bl Time Life Pictures/NASA/The LIFE Picture Collection/Getty Images, br Universal History Archive/UIG via Getty Images; 126 t NASA/JPL-Caltech/SwRI/JunoCam, b NASA/JPL-Caltech/SwRI/MSSS/Kevin M. Gill; 127 t NASA/JPL-Caltech/SwRI/MSSS/Gabriel Fiset, b NASA/JPL-Caltech/SwRI/MSSS/Kevin M. Gill; 128 Atlantic-Press/ullstein bild via Getty Images; 129 t Kean Collection/Getty Images, m NASA/JPL, b DAVID A. KRING / SCIENCE PHOTO LIBRARY; 130 NASA / SCIENCE PHOTO LIBRARY; 131 HarperCollins; 132 NASA/JPL-Caltech; 133 tl and ml NASA/JPL-Caltech/UCLAMPS/DLR/IDA, bl NASA/JPL-Caltech/UCLA/MPS/DLR/IDA/PSI, br NASA/JPL-Caltech/Hap McSween (University of Tennessee), and Andrew Beck and Tim McCoy (Smithsonian Institution); 134 t NASA/JPL-Caltech/UCLA/MPS/DLR/IDA/UMD, b Redrawn from NASA/JPL-Caltech/UCLA/MPS/DLR/IDA; 135 NASA/JPL-Caltech/ASU; 136 NASA / SCIENCE PHOTO LIBRARY; 137 t GEMINI OBSERVATORY / AURA / NSF / NAOJ / JPL-CALTECH / NASA / SCIENCE PHOTO LIBRARY, b NASA / SCIENCE PHOTO LIBRARY; 138 D'July/Shutterstock; 139 NASA/JPL-Caltech/UCLA/MPS/DLR/IDA; 140 NASA/JPL-Caltech/UCLA/MPS/DLR/IDA; 141 ciud/Shutterstock; 142 NASA/JPL-Caltech/UCLA/MPS/DLR/IDA; 143 NASA/JPL-Caltech/UCLA/MPS/DLR/IDA/PSI; 144 l NASA/JPL-Caltech, br Julie Thurston Photography / Getty; 145 t Caltech/Palomar, b HarperCollins; 146 NASA/JPL/University of Arizona; 148 U.S. GEOLOGICAL SURVEY / SCIENCE SOURCE / SCIENCE PHOTO LIBRARY; 149 t USGS Astrogeology Science Center/Wheaton/NASA/JPL-Caltech; 150 US GEOLOGICAL SURVEY / SCIENCE PHOTO LIBRARY; 151 tl NASA/JPL, bl NASA / SCIENCE PHOTO LIBRARY, r US GEOLOGICAL SURVEY / NASA / SCIENCE PHOTO LIBRARY; 152 NASA/JPL-Caltech/SwRI/MSSS/Gerald Eichstadt/Sean Doran; 153 NASA/JPL/Space Science Institute; 154 Tl NASA/JPL-Caltech/SwRI/MSSS/Gerald Eichstad/Sean Doran; 156 NASA/JPL-Caltech/SwRI/MSSS/Gerald Eichstadt/Sean Doran; 157 t NASA/JSC; 158 NASA/JPL/USGS; 159 NASA/JSC; 160 t NASA/ARC, b DAVID JEWITT & JANE LUU / SCIENCE PHOTO LIBRARY; 161 t NASA/ARC, b Oxford Science Archive/Print Collector/Getty Images; 162 NASA/JSC; 163 l HarperCollins, r D. VAN RAVENSWAAY / SCIENCE PHOTO LIBRARY; 164 GEOLOGICAL SURVEY OF CANADA / SCIENCE PHOTO LIBRARY; 165 t NASA/JPL-Caltech, bl The Natural History Museum / Alamy Stock Photo, br Colin Waters / Alamy Stock Photo; 166 HARRINGTON ET AL. / SCIENCE PHOTO LIBRARY; 167 NASA/JPL-Caltech/SwRI/MSSS/Betsy Asher Hall/Gervasio Robles; 169 Shutterstock; 170 NASA/JPL/Space Science Institute; 171 NASA/JPL-Caltech/Space Science Institute/Kevin M. Gill; 172 NASA/JPL-Caltech; 173 HarperCollins; 174 NASA/JPL-Caltech; 175 t NASA/JPL-Caltech/Space Science Institute, b NASA/JPL-Caltech/Harvard-Smithsonian CfA; 176 NASA/JPL-Caltech/SSI; 178 NASA / SCIENCE PHOTO LIBRARY; 179 t NEW YORK PUBLIC LIBRARY / SCIENCE PHOTO LIBRARY, bl and br ROYAL ASTRONOMICAL SOCIETY / SCIENCE PHOTO LIBRARY; 180 NASA; 181 tl NASA/JPL-Caltech/Space Science Institute, tr NASA/JPL-Caltech/Space Science Institute, mr NASA/JPL/Space Science Institute, b NASA Ames; 182 tl NASA/JPL-Caltech/SSI/QMUL, ml NASA/JPL/Space Science Institute, b NASA/JPL/Space Science Institute; 183 t NASA/JPL; 184 NASA / JPL-CALTECH / SPACE SCIENCE INSTITUTE / SCIENCE PHOTO LIBRARY; 185 NASA/JPL-Caltech/Space Science Institute/Hampton University; 186 NASA/JPL/Space Science Institute; 187 NASA/GSFC/Landsat; 188 HarperCollins; 189 t John Cancalosi / naturepl.com, b imageBROKER / Alamy Stock Photo; 190 NASA / JPL-CALTECH / SPACE SCIENCE INSTITUTE / SCIENCE PHOTO LIBRARY; 191 MARK GARLICK / SCIENCE PHOTO LIBRARY; 192 NASA / JPL-CALTECH / SCIENCE PHOTO LIBRARY; 194 t HarperCollins, b NASA/JPL-Caltech/Space Science Institute; 195 l NASA/JPL/KSC, tr NASA, br NASA/JPL-Caltech; 196 t NASA/JPL-Caltech/GSFC/SSI/ESO/IRTF/ESA, b NASA/JPL-Caltech/Space Science Institute; 197 t NASA/JPL-Caltech/Space Science Institute; 198 NASA/JPL-Caltech/SSI; 199 t NASA/JPL-Caltech/SSI/Hampton, b NASA/JPL-Caltech/Space Science Institute; 200 JAE C. HONG/AFP/Getty Images; 201 NASA/JPL-Caltech; 202 NASA/JPL/Space Science Institute; 203 NASA/JPL/Space Science Institute; 204 NASA/JPL; 206 NASA / SCIENCE PHOTO LIBRARY; 207 t NASA, b SSPL/Getty Images; 208 NASA / JPL-CALTECH / SPACE SCIENCE INSTITUTE / SCIENCE PHOTO LIBRARY; 209 tl and tr NASA/JPL-Caltech/Space Science Institute, b NASA / JPL / SPACE SCIENCE INSTITUTE / SCIENCE PHOTO LIBRARY; 210 NASA/JPL-Caltech/Space Science Institute; 211 NASA / JPL / SPACE SCIENCE INSTITUTE / SCIENCE PHOTO LIBRARY; 212 NASA/JPL-Caltech/University of Nantes/University of Arizona; 213 l NASA, ESA, JPL, SSI and Cassini Imaging Team, r NASA/JPL/Space Science Institute; 214 PHAS/UIG via Getty Images; 215 NASA/JSC; 216 NASA / JPL-CALTECH / SPACE SCIENCE INSTITUTE / SCIENCE PHOTO LIBRARY; 217 t HarperCollins, ml NASA / JPL-CALTECH / SPACE SCIENCE INSTITUTE / SCIENCE PHOTO LIBRARY, mr NASA/JPL/Space Science Institute, b NASA/JPL/Space Science Institute; 218 NASA/ESA/IPGP/Labex UnivEarthS/University Paris Diderot; 219 HarperCollins; 220 l EUROPEAN SPACE AGENCY / AEROSPATIALE / SCIENCE PHOTO LIBRARY, tr EUROPEAN SPACE AGENCY / SCIENCE PHOTO LIBRARY, br NASA/KSC; 221 t ESA, b NASA/JPL/ESA/University of Arizona; 222 t NASA/JPL/ESA/University of Arizona, b NASA/JPL-Caltech/ASI/Cornell; 223 NASA/JPL-Caltech, Illustrations by Joby Harris; 224 l NASA/JPL-Caltech/Space Science Institute, r CRISTINA PEDRAZZINI / SCIENCE PHOTO LIBRARY; 225 Universal History Archive/UIG via Getty Images; 226 tl HarperCollins, bl and r NASA/JPL-Caltech/Space Science Institute; 227 t NASA/JPL-Caltech/SSI/PSI, b NOAA PMEL VENTS PROGRAM / SCIENCE PHOTO LIBRARY; 228 NASA / JPL-CALTECH / SPACE SCIENCE INSTITUTE / SCIENCE PHOTO LIBRARY; 229 NASA/JPL-Caltech/Space Science Institute; 231 Shutterstock; 232 Two Micron All-Sky Survey; 233 t Christophel Fine Art/UIG via Getty Images, b NASA, ESA, and the Hubble Heritage Team (STScI/AURA); 234 Space Frontiers/Archive Photos/Getty Images; 235 NASA/GSFC; 236 NASA/JPL-Caltech; 237 HarperCollins; 238 NASA Photo / Alamy Stock Photo; 239 NASA/JPL-Caltech, Illustrations by Invisible Creature; 240 JULIAN BAUM / SCIENCE PHOTO LIBRARY; 241 MPI/Getty Images; 242 MARK GARLICK / SCIENCE PHOTO LIBRARY; 243 tl NASA/JPL, tr CORBIS/Corbis via Getty Images, mr Time Life Pictures/Jet Propulsion Laboratory/NASA/The LIFE Images Collection/Getty Images, bl NASA/JPL, br NASA / ESA / STSCI / E.KARKOSCHKA, U.ARIZONA / SCIENCE PHOTO LIBRARY; 244 l DEA / C.BEVILACQUA/De Agostini/Getty Images, r NASA/JPL-Caltech; 245 AUSCAPE / UIG / SCIENCE PHOTO LIBRARY; 246 l NASA/JPL, r NASA / SCIENCE PHOTO LIBRARY; 247 t SCIENCE SOURCE / SCIENCE PHOTO LIBRARY, b NASA / SCIENCE PHOTO LIBRARY; 248 t NASA / SCIENCE PHOTO LIBRARY, m NASA/ARC, b NASA/JPL; 249 NASA / ESA / STSCI / E.KARKOSCHKA, U.ARIZONA / SCIENCE PHOTO LIBRARY; 250 US GEOLOGICAL SURVEY / SCIENCE PHOTO LIBRARY; 251 tl NASA/JPL, tr, bl and br NASA / SCIENCE PHOTO LIBRARY; 252 NASA/ARC; 253 t Universal History Archive/UIG via Getty Images, b NASA/JPL; 254 t ANN RONAN PICTURE LIBRARY / HERITAGE IMAGES / SCIENCE PHOTO LIBRARY, b NASA/ARC; 255 t ROYAL ASTRONOMICAL SOCIETY / SCIENCE PHOTO LIBRARY, b NASA/ARC; 256 MARK GARLICK / SCIENCE PHOTO LIBRARY; 257 tl and tr NASA/JPL, bl and br NASA/JPL/STScI; 258 DETLEV VAN RAVENSWAAY / SCIENCE PHOTO LIBRARY; 259 LOUISE MURRAY / SCIENCE PHOTO LIBRARY; 260 t NASA/JPL, b NASA/JPL-Caltech/Lunar & Planetary Institute; 261 t NASA/JPL, b NASA/JPL/USGS; 262 JOHN R. FOSTER / SCIENCE PHOTO LIBRARY; 263 NASA/JPL; 264 TAKE 27 LTD / SCIENCE PHOTO LIBRARY; 266 t Bettmann / Getty, b SSPL/Getty Images; 267 t U.S. Naval Observatory, b Universal History Archive/UIG via Getty Images; 268 NASA / JHUAPL / SWRI / SCIENCE PHOTO LIBRARY; 269 BRUCE WEAVER/AFP/Getty Images; 270 t HarperCollins, m NASA/Johns Hopkins University Applied Physics Laboratory/Southwest Research Institute, b NASA/Caltech; 271 l MARK GARLICK / SCIENCE PHOTO LIBRARY, r JOHN R. FOSTER / SCIENCE PHOTO LIBRARY; 272 NASA / JHUAPL / SWRI / SCIENCE PHOTO LIBRARY; 273 t NASA/Johns Hopkins University Applied Physics Laboratory/Southwest Research Institute, bl NASA/Johns Hopkins University Applied Physics Laboratory/Southwest Research Institute, br NASA, ESA, M. SHOWALTER (SETI INST.), G. BACON (STSCI) / SCIENCE PHOTO LIBRARY; 274 t SCOTT CAMAZINE / SCIENCE PHOTO LIBRARY, b NASA/Johns Hopkins University Applied Physics Laboratory/Southwest Research Institute; 275 t NASA/Johns Hopkins University Applied Physics Laboratory/Southwest Research Institute; 276 NASA/JHUAPL/SwRI; 277 tl, tr, bl NASA/Johns Hopkins University Applied Physics Laboratory/Southwest Research Institute, br NASA/JHUAPL/SwRI; 278 t NASA, ESA, SWRI, JHU / APL, AND THE NEW HORIZONS KBO SEARCH TEAM / SCIENCE PHOTO LIBRARY, m NASA / JOHNS HOPKINS UNIVERSITY APPLIED PHYSICS LABORATORY / SOUTHWEST RESEARCH INSTITUTE / SCIENCE PHOTO LIBRARY, b HarperCollins, redrawn from NASA/Johns Hopkins University Applied Physics Laboratory/Southwest Research Institute; 279 t NASA/Johns Hopkins Applied Physics Laboratory/Southwest Research Institute, National Optical Astronomy Observatory, b WALTER MYERS / SCIENCE PHOTO LIBRARY; 280 NASA / JHUAPL / SWRI / SCIENCE PHOTO LIBRARY